Biosciences
on the Internet

Biosciences on the Internet

A Student's Guide

Georges Dussart

Canterbury Christ Church, University College, UK

JOHN WILEY & SONS, LTD

Copyright © 2002 by John Wiley & Sons Ltd,
Baffins Lane, Chichester,
West Sussex PO19 1UD, England

National   01243 779777
International (+44) 1243 779777

e-mail (for orders and customer service enquiries): cs-books@wiley.co.uk

Visit our Home Page on http://www.wileyeurope.com
or
http://www.wiley.com

Other Wiley Editorial Offices

John Wiley & Sons, Inc., 605 Third Avenue,
New York, NY 10158-0012, USA

Wiley-VCH Verlag GmbH, Pappelallee 3,
D-69469 Weinheim, Germany

John Wiley & Sons (Australia) Ltd, 33 Park Road, Milton,
Queensland 4064, Australia

John Wiley & Sons (Asia) Pte Ltd, 2 Clementi Loop #02-01,
Jin Xing Distripark, Singapore 0512

John Wiley & Sons (Canada), Ltd, 22 Worcester Road,
Rexdale, Ontario M9W 1L1, Canada

Library of Congress Cataloging-in-Publication Data
(applied for)

British Library Cataloguing in Publication Data

A catalogue record for this book is available from the British Library

ISBN 0-471 49842 4 (Paperback)

Typeset by Dobbie Typesetting Limited, Tavistock, Devon
Printed and bound in Great Britain by Biddles Ltd, Guildford and King's Lynn
This book is printed on acid-free paper responsibly manufactured from sustainable forestry,
in which at least two trees are planted for each one used for paper production.

Contents

Preface

While bringing advantages of depth, scope and communication, the ease with which information, ranging from excellent to poor quality, can be obtained from the internet presents problems of both *quality and quantity. Copyright* is also a serious issue. In journals, periodicals and books, the reader can be sure that the work has been critically evaluated before publication. Although there are cases of extreme views on the internet, it is usually easy to recognise these for what they are. However, there may be more subtle situations where a naive student might not recognise that information is unreliable, or that a particular line is being adopted by the author.

Teachers are likely to find it difficult to keep up with the wealth of rapidly changing information and the student undertaking research may be substantially alone in 'hyperspace'. Peter Boyce has suggested that in future the whole internet may be the electronic journal; sophisticated search programs could mean that there is only one journal, and it is the internet.

Students need to be aware of the need for care when they are loose on the internet. Even primary school children (5–11 y) are being introduced to the internet and education for prudence therefore needs to begin at a young age. The development of a *prudent* attitude should be guided and monitored by both parents and teachers and this book represents a contribution to this effort.

The aims of the book are to:

- introduce *bioscience students* to procedures for efficiently using the internet;

- review the benefits and problems of internet use, including issues of copyright and plagiarism;

- review a range of bioscience sites. These have been classified on the basis of the England and Wales advanced-level curriculum studied by 16–18-year-old biology students, but many other bioscience sites have been included.

The book is therefore primarily aimed at *senior secondary school students and first year undergraduates* but should be useful for teachers of biosciences at all levels. Some sections are specifically addressed to teachers (e.g. 3.4).

As an experienced university tutor, I am aware that many students lack *essential skills* in some areas of their performance. Consequently, the brief was extended to include aspects such as *essay writing, poster presentations and organisation of information*. These skills do not directly underpin use of the internet but will almost certainly be employed in relation to the downloading of information.

The Higher Education Funding Council in the UK is now specifying the nature of study skills for university programmes. *Generic patterns of skills* are appearing. While the sources used in a literature search might differ from subject to subject, the nature of the investigation may transcend disciplines. For example, a historian might investigate county court records and a biologist might investigate the raw data held by the Environmental Agency. However, the methodology in both disciplines uses a similar kind of logic. Consequently, this book was not intended to be a directory of methods which are unique to biosciences but was intended to offer *common-sense advice* on how to use the internet in a biosciences context.

Although the book addresses generic issues, it is written by a particular kind of biologist – in this case an ecologist. It therefore presents my subjective view as an experienced university teacher. Nevertheless, I would like to think that a molecular biologist might have written a similar kind of book, though the examples would undoubtedly differ. Although I have tried to focus on biological examples throughout the text, hopefully the book will be useful to students from a range of disciplines.

Of necessity, most of the images and procedures have to relate to a particular system. In this case, I have tended to use Netscape rather than Internet Explorer. Currently, the former has more useful features, is easier to use and is more sophisticated. I have also occasionally referred to Windows-based software packages such as the word processor package Microsoft Word. Apple users should find that the advice is general enough to apply to Apple systems too.

The book has three parts. Part 1 concerns *basic principles* of using the internet, including issues of *plagiarism* ('copying'), and *evaluation of quality* in web sites and *managing files*. In Part 2, the reader is guided through some *typical searches*. I have tried to make this section as readable as possible; ideally it will entertain as well as be instructive. Part 3 is an *annotated list of web sites*; these are websites that appear to contain reliable information and appear to be relatively stable in time. To check the latter, these sites were visited a number of times over a period of months. Sites which were not stable were dropped. Part 3 also contains a list of subjects commonly used by national bodies in the British educational system which examine students aged 16–18 years (A levels), just prior to attending university. Each of these subjects has been linked to an appropriate web site. Here, I have tried, perhaps with limited success, to select sites which were appropriate to the level of the students concerned.

To try to make what could be a dry subject more readable, I have used the personal pronoun ('I') where there is an action, such as a search, which I have undertaken personally. The second person ('you') has been used when offering advice on what the reader might do in a particular situation. In general, I have assumed that readers will have access to a computer and know the basics of switching on, logging on and word processing. Just in case this is an over-assumption, the appendix contains a brief description of how to set up the elementary hardware of a home computer system.

ACKNOWLEDGEMENTS

I would like to thank Dr Mariann Rand-Weaver of Brunel University, Dr Sue Williams of the University of Hertfordshire, Dr Ruth O'Riordin of University College Dublin, Dr Geoff Lovell of Kingston University, and Dr Nicholas Watinough of the University of East Anglia, for giving generously of their time in reading the manuscript. I would also like to thank Rachel Ballard of Wiley for her help and would particularly like to thank Nicky McGirr of Wiley for her enthusiastic and diplomatic support throughout the process of writing the book. Claire Beverley, Philip Buckley, John Badmin, Ray Calleja, Peter Gilchrist, Dr Mike Nicholls and Dr Jackie Trigwell are also thanked for their contributions. Finally, I would like to thank my wife for her forbearance; without her good humour, this book could not have been written.

GBJD
August 2001

Permissions and credits

Figure I.5, E. Lissimore, Nelson-Thorne;

Figure I.4, G. Goodwin, Macmillan;

Figure I.6, C. McBride, McGraw Hill;

Figure I.10, 5.1, 15.6, 15.7, C. McCaffrey, Google, Inc;

Figure 2.6, Candace Moses, BBC;

Figure 2.9, ©Crown Meteorological Offfice, UK;

Figure 2.10, K. Dalkowski, Environment Canada. Reproduced with the permission of the Minister of Public Works and Government Services Canada, 2001;

Figure 3.4, Prof. Dr. H. Mehlhorn, Germany;

Figure 4.1, D. Rossie for ATSDR, CDC;

Figure 4.2–4.6, Biobest n.v., Belgium;

Figure 4.11, F. Willis, Oxford University Press;

Figure 4.12, L. Millhouse, Natural History Museum;

Figure 4.13, A. Gibbins, Royal Society;

Figure 4.14, A. Bourton, New Scientist;

Figure 4.15, A. Grimwade, The Scientist;

Figure 4.16, P. Liu, Biomednet;

Figure 4.17, S. McGinnis, National Center of Biotechnology Information (PubMed);

Figure 4.18, J. Tamames;

Figure 4.19–4.20, C. Mackenzie, ISI;

Figure 4.21, L. van den Dolder, Sinuaer Associates;

Figure 4.22, C. McBride, McGraw-Hill;

Figure 4.23, M. Street, Wiley;

Figure 4.24, A. Brown, Highwire;

Figure 5.2, D. Reid, Ask Jeeves;

Figure 6.4, P. Varney, ©Topica;

Figure 7.1, M. Davies, Malacological Society of London;

Figure 7.2, L. Gauthier, CITES;

Figure 7.5, ©Sylvie Lapègue, E. Diaz, A. Imela, S. Launey, C. Ledu, P. Boudry, Y. Naciri-Graven, F. Bonhomme;

Figure 10.1–10.8, Dr. Alan Cann, UK;

Figure 12.1–12.2, Figure 15.13–15.15, B. Starkie, WeberShandwick;

Figure 12.3–12.7, J. Precious, reproduced by permission of The Stationery Office Limited;

Figure 13.2–13.5, Europa;

Figure 14.1, D. Eck, Field Museum, US;

Figure 15.1, J. Macpherson, AltaVista;

Figure 15.2–15.5, M. Freeman, INWR;

Figure 15.8–15.9, Brian Lucas, GridA, Norway;

Figure 15.10-15.12, Halvard Wensel, Norwegian Ministry of Fisheries;
Figure 15.16, ©Scott Polar Research Institute, University of Cambridge;
Figure 15.17-15.21, G. Donovan, IWC;
Figure 16.1-16.4, R. Wilson, ICRF;
Figure 17.1, S. Gomez, ©Institute of Biology;
Figure 17.2, R. Bourgeois, IPCC;
Figure 17.3, A. Knee, IUCN;
Figure 17.4, J. Slotta, WISE;
Figure 17.5, ©G. Chapelle, Carcinologie, IRScNB;
Figure 17.6, T. Middleton, Network 2002, UNED Forum;
Figure 17.7, Sustainable Development International.

Introduction

In his history of the origins of the internet, John Naughton (1999) quotes the poet Yeats, saying 'a terrible beauty is born'. As it is currently organised, the newborn internet is a global medium for free speech, offering a voice by which *any citizen* can address potentially *all citizens*. However, it also gives the most dangerous elements of society, including governments, an information distribution system which can be abused. And the rate of use of the internet is rising exponentially (Fig. I.1). Four years ago, as a university teacher, I only rarely received assignments with references to material on the internet. Students' reference lists would be directed towards materials in our university library, or in local libraries. Now, frequently, the reference list has been compiled exclusively from the internet. Four years ago, one of my relatives (60 years old) claimed that she would never buy, or use, a computer. Now, she maintains her family-run bed and breakfast business via advertising and reservations made over the internet.

So, how might this developing information system affect the biologist? Figure I.2 attempts to portray the geography of the traffic over the internet. Biologists need to address the issue of the increasing use of internet resources by students at all levels, ranging from primary school to university. Networked electronic media such as the internet have a number of advantages. Students can search vast fields of material in great depth and can also communicate with others who have similar interests. Material can be accessed that would otherwise be unavailable. For example, expedition reports can be produced while an expedition is actually taking place, as shown by Liv Arnesen and Ann Bancroft who attempted to cross Antarctica:

http://www.yourexpedition.com 4/4/01–5/7/01*

A more dubious advantage is that students can easily, intentionally or unintentionally, make copies without appropriate attribution of authorship.

---

*Throughout this book, web addresses are followed by the first and last dates on which the web site was visited. The length of this period gives some idea of the stability and permanence of the site.

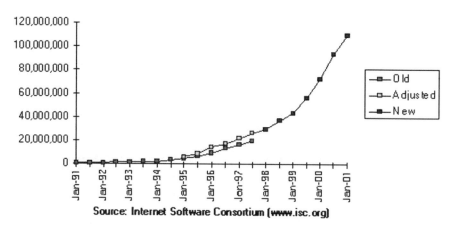

Figure I.1 Exponential increase in the number of computers which host web information. Source: Internet Software Consortium http://www.isc.org/ds/hosts.html 5/4/01–5/7/01

WHAT IS THE INTERNET?

In this text, the 'web' and the 'internet' are used interchangeably. The 'world wide web' is a particular domain of the internet.

The internet is the total of all the computers in the world which are

- *connected together* and

- *exchanging information* with each other.

It is a dynamic, growing structure. As more computers are added to the system it grows ever more extensive and comprehensive. The internet has two major capabilities. Firstly, it can be used as a reservoir, or *repository*, of knowledge. Secondly, it can be *searched* at great speed. However, this speed means that the searcher can rapidly be supplied with huge amounts of material to sift through, much of it being irrelevant. Deciding how to search efficiently, and how to sift through large amounts of material, are important skills for anyone who ventures out onto the internet.

Through the internet you can currently:

- *search* for information anywhere in the world using key words and phrases

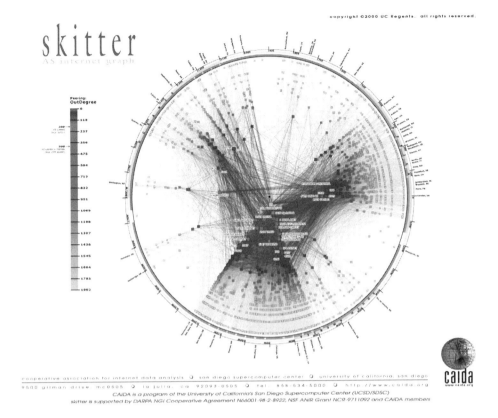

Fig. I.2 A conceptual map of internet traffic across the globe at http://www.caida.org/ analysis/topology/as_core_network/AS_Network.xml 4/4/01–5/7/01. Source: Internet Software Consortium http://www.isc.org/ds/hosts.html

- *copy* text, sound and images
- *send and receive* text messages by email
- to these messages, you can *attach*
    - ⇒ pages of words (text files)
    - ⇒ pictures (image files)
    - ⇒ moving images (videoclips)
    - ⇒ sound files
- *listen* to radio and *watch* television

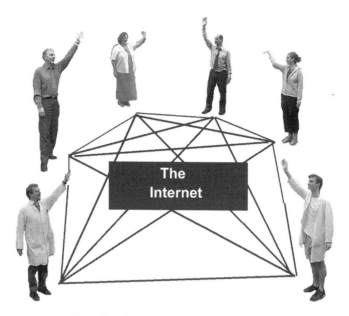

Fig. I.3   The internet is literally a network

- *chat* with others on-line

- *investigate* goods and services – for example check your bank account

One of the *major attributes* of the internet is that information can be found easily and quickly, and often at low cost. The digital electronic processes which underpin the system allow you to search efficiently for the information you need without physically going to where the information is located. You go to the source of the information as a *virtual visitor* rather than a real visitor. The information is held on a web site. The web site can include one, several or even thousands of pages.

However, the internet is not perfect. Bruce Royan (2000) has likened it to a car boot sale rather than a digital network and Tim Berners Lee (2000), who invented the language used on the internet, said the web will always be 'a little bit broken'.

One of the *major defects* of the internet is that it offers a new way to steal, cheat, confuse and damage people. It is not perfect and people must be careful.

So what are the issues faced by a student who wants to investigate biosciences via the internet?

There are four challenges:

1 Arthur C. Clarke (2000) has said that 'getting information from the internet is like getting a glass of water from Niagara Falls'. How *do* you get sufficient information without being swept away?

2 How do you search *efficiently*?

3 How do you make sure that what you get is *reliable and relevant*?

4 How do you present what you have got *without infringing copyright or indulging in plagiarism*?

In the following pages, it is hoped that the reader will be guided to solutions to these challenges.

THE PRINCIPLES AND SOME BASIC TERMINOLOGY

Hardware

The basics of setting up a computer, printer and monitor, etc. are described in the Appendix.

Multiple media and multimedia

'*Multiple media*' are the materials of audiovisual presentations – for example, diapositive slide projector, overhead projector, data projector, whiteboards and flip charts.

'*Multimedia*' means a range of electronic materials collectively available via the same computer. For example, a multimedia computer is usually equipped with a CD-ROM or DVD drive, soundcard, speakers, and image processing capability so that moving images can be viewed. A multimedia delivery probably includes a mix of different forms of presentation which will all be linked together. For example, in a single program there could be a mixture of text, images, sound and video.

Software

Books and web sites – similarities and differences

The basic structure of a biology textbook comprises:

- front and back covers

- a preface saying what the book is about and for whom it is intended, and acknowledgements

- a contents list

- an introduction

- chapters of text

- references (which might come at the end of each chapter)

- an alphabetical index

Because all books have this standardised arrangement, it means that they are simple to use and it is easy to find information. The reader is usually interested in what is *in* the book rather than the aesthetic arrangements of the book. In general, the reader wants a simple, clear presentation, so that the book makes learning easy, rather than making it difficult.

Early biology textbooks had a densely worded text in small font with complex sentences and cluttered drawings (for example, Murray, 1952; Fig. I.4). A major milestone was the secondary-school A-level text by M. V. Roberts, *Biology - a Functional Approach* published by Nelson Thornes in 1972. The comparative, functional approach was new and refreshing, and the figures were clear and informative (Fig. I.5). Roberts made a special effort to write simple sentences aimed at the exact age group of his audience. This book attracted a whole generation of students into biology. In the 1980s and 1990s, a new suite of largely American biology texts were published with a clear organisation, coloured illustrations and lucid sentence constructions (for example, Guttman, 1999; Fig. I.6). An innovation was to supply banks of *self-test questions* so that students could take some responsibility for their own learning. Like Roberts, these books attempted to attract students into the subject, to motivate them, to guide their learning and, ultimately, to pass them on to new, higher learning experiences. These books now have companion web sites which offer many kinds of student and tutor support materials (see section 4.6.8).

Now, on the internet there are similar kinds of resources, often made available by authors who subscribe to the view that *good information should be freely available*.

Web pages usually have the same *objectives* as a book. The web author hopes to attract the reader to visit the site, then to offer some useful information and subsequently to pass them on to other sources of relevant information. However, just as a book can be too big to handle easily, or be

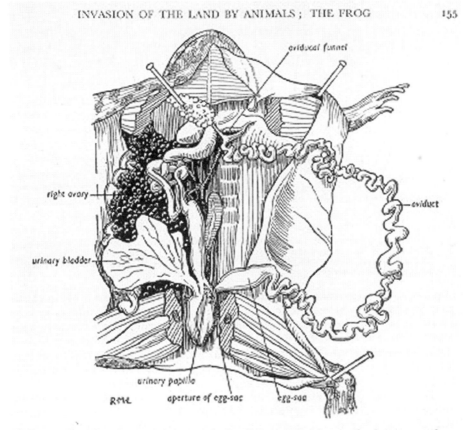

INVASION OF THE LAND BY ANIMALS; THE FROG   155

FIG. 109. The urino-genital organs of a female frog. The left ovary has been removed.

open into the cloaca in front of the ureters. Their walls secrete the jelly-like material in which frog eggs are found embedded, and at the posterior end each is enlarged as an *egg-sac* in which the eggs are stored before deposition.

The breeding habits are described on p. 178.

Kidneys and urino-genital ducts in vertebrates generally

The pro-, meso- and meta-nephroi and their ducts. The kidneys of vertebrates develop in a curious way. There is formed in the early embryo a pair of structures called *pronephroi* (sing. *pronephros*). The pronephroi are formed dorso-laterally at the anterior end of the trunk, just outside the peritoneal cavity, and each consists of two or three tubules arranged segmentally (one pair to a segment), each tubule

Figure I.4   Page from Murray

Fish
single circulation

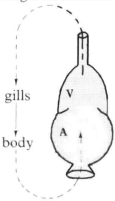

gills

body

Amphibian
double circulation with
partially divided heart

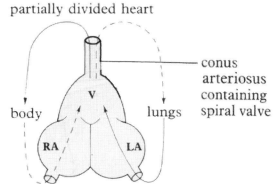

body   lungs

conus
arteriosus
containing
spiral valve

Mammal
double circulation with
completely divided heart

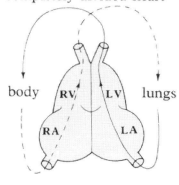

body   lungs

Octopus
separate hearts for body and gills

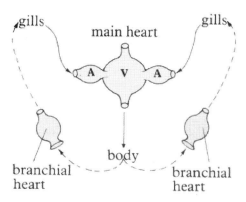

gills   gills

main heart

body

branchial
heart

branchial
heart

Figure I.5 Single and double circulations. The hearts are diagrammatic. In the amphibian and mammal they have been deflected forward so that the atria appear to be behind the ventricles. A, atrium; V, ventricle; R, right; L, left

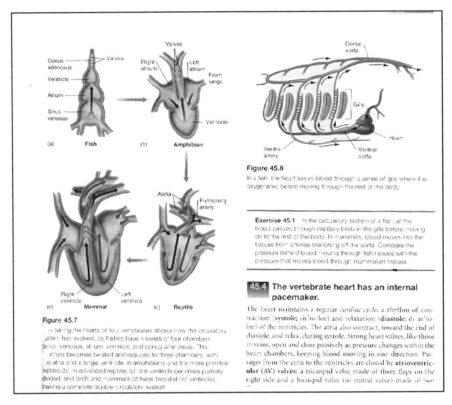

Figure 45.8

In a fish, the heart forces blood through a series of gills where it is oxygenated before moving through the rest of the body.

Exercise 45.1 In the circulatory system of a fish, all the blood passes through capillary beds in the gills before moving on to the rest of the body. In mammals, blood moves into the tissues from arteries branching off the aorta. Compare the pressure behind blood moving through fish tissues with the pressure that moves blood through mammalian tissues.

45.4 The vertebrate heart has an internal pacemaker.

The heart maintains a regular *cardiac cycle*, a rhythm of contraction (systole; sis'to-lee) and relaxation (diastole; di as'to-lee) of the ventricles. The atria also contract, toward the end of diastole, and relax, during systole. Strong heart valves, like those in veins, open and close passively as pressure changes within the heart chambers, keeping blood moving in one direction. Passages from the atria to the ventricles are closed by **atrioventricular (AV) valves**: a tricuspid valve made of three flaps on the right side and a bicuspid valve (or mitral valve) made of two

Figure 45.7

Comparing the hearts of four vertebrates shows how the circulatory system has evolved. (a) Fishes have a series of four chambers: sinus venosus, atrium, ventricle, and conus arteriosus. This structure becomes twisted and reduced to three chambers, with two atria and a single ventricle, in amphibians and the more primitive reptiles (b). In advanced reptiles (c) the ventricle becomes partially divided, and birds and mammals (d) have two distinct ventricles, making a complete double-circulatory system.

Figure I.6 Page from Guttman

prone to gimmicks (for example, cartoons), or have too dense a text, the organisation of a web page can come between the intentions of the author and the needs of the reader.

The usual, general structure of a web site includes

1 A *homepage* which illustrates the general purposes of the site

2 A *clickable index* to the pages which make up the site

3 A *final section* which describes who wrote the pages and how they can be contacted.

A web site usually includes two frames. One frame contains information that you might want to refer to at any time while you are visiting the site. The other frame contains a page which is linked to other pages. The whole structure can be represented as an *inverted tree*, or *dendrogram*.

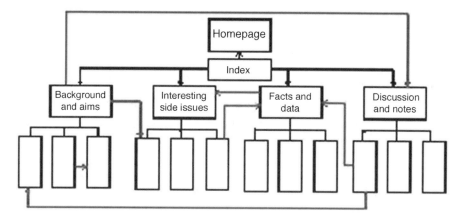

Figure I.7 Organisation of a web site

In Figure I.7, the main links are shown but there might be other hypertext links which the author thinks would be useful to the reader.

Good web sites should have at least:

- an *index* page or *homepage* which has hypertext links directing the reader to more detailed pages on the web site. Hypertext links are words or images which, when selected with the cursor, will take the reader to another web page;

- *additional pages* of information;

- *links* to other web sites which might be of interest;

- a *search engine* for the site. A search engine is a program which searches the internet for selected key words;

- a *copyright notice*;

- a *disclaimer*;

- a *contact address* and the email address of the person who constructed the web pages.

The internet comprises a *client–server* system. The client asks, and the server supplies.

Web sites are held on computers called '*servers*'. These computers are owned and controlled by *internet service providers* (ISPs). Typical ISPs include AOL (America OnLine), Compuserve and Microsoft Network (MSN) (Fig. I.8).

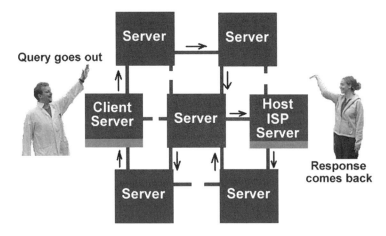

Figure I.8 The client-server relationship across the internet

If you want to have your own web site, you create it using special languages such as *html* or Java and then upload it to the ISP. ISPs usually make a charge for this.

If you are working in an educational context, the institution will have its own server which can host your web page. People (clients) who are interested in what you have to offer on your web page can read it as virtual visitors. It is useful to have a *counter* on your web page so that you can see how many visitors you have received. Such counters can be downloaded free from the internet (e.g. http://more.bcentral.com/fastcounter 6/5/00–23/7/01).

Your web site will be visited and examined by software programs called *search engines*. Some search engines will examine the whole web site for keywords; others will refer to the first part of the web site program, called the *metafile*, which contains information describing your web site.

Sitting at your computer, you cannot correspond directly with the servers of the internet. You need to have a piece of software called a '*browser*' (Figure I.9). Browsers come ready installed with the software on your computer or you can download browsers from the internet. Obviously there is a Catch 22 here since *you need a browser to connect to the internet to get a browser*.

Many browsers are available but the three most common current examples are Netscape, Internet Explorer and Opera. All software is continually being upgraded and you sometimes have to pay for the newest *upgrades*. However, quite serviceable early versions are available as free downloads. Personally, I prefer working with Netscape. It allows you to see which pages you will print, seems to have no problems with downloading

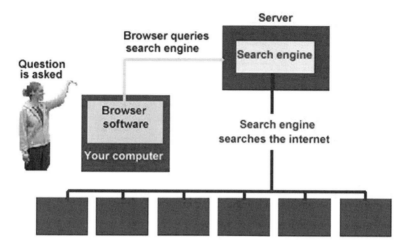

Figure I.9   How the browser, a search engine and the internet relate to each other

images and has no irritating split screen. Not all of this is true of the other browsers.

- Netscape can be downloaded *free* from

    http://home.netscape.com/download 6/5/00–23/7/01

- Internet Explorer can be downloaded free from

    http://www.microsoft.com/windows/ie/default.htm 6/5/00–23/7/01

- Opera is a relatively simple and therefore fast browser which make little demand on the hard disc space. It can be downloaded free from

    http://www.operasoft.com/download 6/5/00–23/7/01

Newsgroups

A newsgroup is an electronic notice board allowing free discussion of a topic by a group of interested people, who are called 'subscribers'. All the newsgroups on the internet comprise *usenet*. It is the computers and software that share news information. Some groups, particularly if they are in some way official, have a moderator who controls what is going on. The newsgroup operates according to an agreed charter and contributors are usually anonymous. Giving personal details can make people vulnerable to personal intrusion in their real, as opposed to cyber, lives, since the information can be read by anyone. In normal conversation, people avoid being offensive by social controls of body language etc. These controls do not exist on the web and a message can appear to be more abrupt or offensive than was intended. Discussions can quickly become unpleasant and abusive (*flaming*). It is possible to try and tone down critical comments by using *emoticons* such as:

$$:\text{-})=I \text{ am happy}$$

$$:\text{-}(=I \text{ am sad}$$

There is a gallery of emoticons at:

http://www.randomhouse.com/features/davebarry/emoticon.html 5/5/99–23/7/01

However, I think the best strategy is to behave as if you were in a local conversation with someone and be reasonably polite and guarded. For example, it is useful to begin every communication with 'Dear...' and finish with 'Best wishes...'. Both are relatively neutral and do not allow the reader to misinterpret the emotions of the writer.

You can get an overview of some of the available biological newsgroups by going to a search engine and typing in 'biology newsgroups'. For example, Fig. I.10 shows a newsgroup conversation concerning *women in biology*.

Software on the net

Different types of software are available to you over the internet. These include:

- *Public domain* – copyright-free so you can do what you like with it
- *Shareware* – free on trial but then should be paid for
- *Freeware* – can be used and distributed but not altered
- *Fully commercial* – has to be licensed and paid for.

Types of data

Files can include maps, music, images, text, software programs and data observations (such as weather readings). Big files are best downloaded using programs called '*file transfer protocols*' (*FTP*), which can take a long

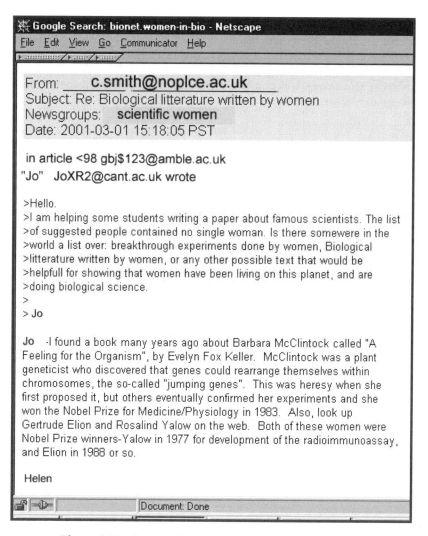

Figure I.10 Extract of a newsgroup on women in biology

time. To make downloading quicker and easier, some files are compressed into *zip files* (i.e. they are zipped up). You then need a zip program such as Winzip (http://www.winzip.com/upgradex.htm 5/10/00–23/7/01) to decompress or unzip the files so that you can look at them. You can also buy cheap software over the web which will help you to make fast loads (for example, http://www.globalscape.com/products/cuteftp/index.shtml 5/10/00–23/7/01).

Intranet and internet

Institutions such as schools, colleges and universities usually have their computers linked together to form a *local area network* (*LAN*) *or intranet*. This intranet can then communicate through its own servers to the internet. Incoming information must pass through a protective *firewall*. Since the link between inter- and intranet needs to carry lots of information, then there is usually a special connection which can carry large amounts of data because of its wide bandwidth. This broadband connection is usually through a particular type of telephone line called an '*ISDN line*' which is open all the time (i.e. on-line).

HISTORICAL CONTEXT AND BACKGROUND TO THE INTERNET

How it Started

The internet began in university research laboratories, where researchers attempted to use computers to communicate with each other. In the beginning it was relatively easy to communicate via computers. Two computers could be connected and could share information via telephone links. However, as more computers joined each other, management of the information became more difficult and also locked up large amounts of computer memory.

A new approach had to be developed and therefore a system of *packet switching* was developed. In this process, a message would be broken into small packets, each of which was sent separately, together with a note of how they had been broken down. At the receiving end, the note could be used to reassemble the message. In this way, millions of small messages can be disassembled, sent and reassembled across the world, every second. Of course, this degree of information handling depended on technological breakthroughs in communications (using lasers and fibre optics), in computer hardware (allowing rapid access to large amounts of computer memory) and in computer software (the development of appropriate programs to handle all this).

The Language

A major step forward was the development of a *universal language* by Tim Berners Lee at CERN (the European Organization for Nuclear Research) which allowed everyone to see the same arrangements of material on their computer screens. This step arguably marked the start of the *world wide web* (*www*). The language which allowed this to be done comprised a few simple rules and was called '*hypertext mark-up language*' (*html*).

The principle of this language is that an instruction is switched on, then switched off by a forward slash character (/). Thus the instruction to print some text in bold would look like this: . . .

<bold> What happens now? </bold>

and would look like this on the computer screen:

What happens now?

With only twelve html commands, it is possible to produce a quite serviceable web page such as that shown in Fig. I.11.

Web searching

Here is a protocol for searching the web for concrete information

Invoke a good search engine such as Google

Before you leave the search engine to visit a promising website, check the webaddress and

ask yourself the following questions !

· Are there clues that the hosts will want money ?
· Do the hosts have a vested interest ?
· Is the site one which won't give you anything concrete e.g. a Society ?
· Are the hosts sufficiently academic ? (in other words, you don't usually want to be looking at the work of other students)

Figure I.11   A simple web page constructed from twelve commands

`<html>`	html indicates that the text will be in this language
`<body bgcolor = "#D9FFFF">`	body identifies the text
`<h2>Web searching</h2>`	bgcolor is background colour (light blue)
`<h3><font color = "#000066">Here is a protocol for searching the web for concrete information </font> </h3>`	h2 = large heading
`<b>Invoke a good search engine such as Google`	h3 = smaller heading
` Before you leave the search engine to visit a promising website, check the webaddress and`	b = make the font bold
`<h3> <font color = "#990000">ask yourself the following questions </font> </h3>`	font color number means a dark blue font (#000066") or a red font (#990000)
`<li>Are there clues that the hosts will want money?`	br = line break
`<li>Do the hosts have a vested interest?`	ul = unordered list
`<li>Is the site one which won't give you anything concrete e.g. a Society?`	li = list
`<li>Are the hosts sufficiently academic? <font size="3">(in other words, you don't usually want to be looking at the work of other students)</font></b>`	font size = "3" means make the font small
` `	
` `	
`<img src = "leaves.jpg"> Leaves of a date palm`	img src means bring in an image, in this case "leaves"
`</body>`	
`</html>`	

Figure I.12 A simple program for a web page, with an explanation of the code

This web page was produced using Notepad, which is a simple word-processing program supplied with all Windows packages. The html code is shown in Fig. I.12. Try to map the commands in the right-hand column to the lines on the left.

A particular advantage of html is the use of *hypertext links*. A hypertext link is a highlighted word which can act as a connection to a file held somewhere else, for example nearby in the same computer, or further afield, such as on another continent. Hypertext links usually look like this: hypertext (underlined and usually blue). So, if this was a web page, when you clicked on the highlighted hypertext word in the previous line, it would take you somewhere.

The Ethos

Throughout the development period in the second half of the twentieth century, there was an atmosphere of cooperation and sharing of

1945	Vannevar Bush has the idea of the web in an article called 'As we may think'
1946	J. Licklider and others develop the idea of an internet, involving a seamless informational relationship between the human and the computer
1961	John McCarthy and Fernando Combato at the Massachusetts Institute of Technology link together four computer terminals
1961–	J. Licklider and R. Taylor working at the Advanced Project Research Agency (ARPA) for the Pentagon are involved in the development of many of the strategic aspects of the internet
1962	Paul Baran at the Rand Corporation envisions and develops a distributed network of hardware which underpins the current internet. Rand was involved with defence contracts at the time but Baran's work was a separate issue
1963	Donald Davies in the UK develops the idea of packet switching
1965	Hypertext is envisaged by T. Holm Nelson
1966	Davies, Bartlett, Scantleby and Wilkinson from the UK read a paper describing a fully functioning internet system to a symposium in Tennessee, which ultimately brings together ARPA and Rand in the USA
1968	D. Englebert demonstrates multiple windows, keyboard with a cord, mouse and graphics interfaces
1969 Oct.	Stanford and UCLA are linked and send the first internet message
1969	ARPA contracts the development of an internet to a company, *Bolt, Beranek & Newman*.
1970s	K. Thompson and D. Ritchie at AT&T develop the UNIX operating system, out of hours, on an obsolete computer. UNIX economically deals with files, task switching and device drivers. It leads to freely available software (called 'open source software'). The idea of free availability multiplies, particularly through the relationships between Apple personal computers and the internet
1973	Leonard Kleinrock asks the first significant question over the international internet – '*Can someone bring my razor back to the USA from the conference I attended in Sussex, UK?*'
1974	Vinton Cerf and Robert Kahn conceptualise sending messages through *Transmission Control Protocols* (TCP)
1977	*Gateways* are developed to bridge electronic mail systems. *Routers* translate the TCP between networks
1985	Internet protocols are incorporated into UNIX computer systems at the University of California, Berkeley, and the internet becomes operational
1990	The html language and hypertext links are developed by Tim Berners Lee at CERN in Switzerland; Nicola Pellow develops the first internet searching program (a browser)
1991 15 Jan.	This is the day the web goes public; personal computers are now generally available
1993	The first major browser, *Mosaic*, is developed by M. Andreesen, serving as a platform for the development of *Netscape* a year later
1991	Linus Torvalds makes the program code for the *Linux* operating system generally available via the internet
1994	Microsoft enters the internet game, very late, with its clumsy browser *Internet Explorer*. However, with its massive marketing power, it begins to take over
1998	*Netscape* is released as open source material
	From here on, the internet shows a phenomenal development

Figure I.13   A chronology of the internet (from Naughton, 1999)

information that is still the prevailing ethos of the internet. There was a general agreement that

> nothing was secret, that problems existed to be solved, that solutions emerged iteratively and that everything that was produced would be in the public domain.
>
> (Naughton, 1999)

This ethos also underpins this book, where the focus is on bioscience information that is freely available, rather than knowledge which has to be paid for or which is otherwise locked away from public view.

Incidentally, the first message to be sent experimentally over the internet was the three-letter word 'log' – after this the system crashed. This was rather prescient for the way systems would perform later.

Who Did What and When in the Development of the Internet

As well as complementing your general knowledge and helping you to get a feel for what the internet is, I present a chronology containing the kind of information which could be useful in quiz games (Fig. I.13). It is invidious to identify individuals where team efforts were involved but there are some key names and dates in the time-line of the internet.

Part 1

Teaching and learning in relation to the internet

1
Approaching the Internet

1.1 INTEGRITY AND LEARNING

1.1.1 Intellectual Honesty

It is the dream of any teacher at any level to have a student who is genuinely motivated. It implies a shared enthusiasm between teacher and student, a real interest in the subject and a desire to learn. Most teachers would seriously dedicate their time to such a student. A student with a desire to learn has the following objectives:

- to increase the capacity of their *intellect*,
- to improve their *understanding*,
- to build on their *skills* so that they become accomplished in their chosen subject.

It is a great compliment when a student chooses a teacher with whom they want to do this and the teacher invariably responds with goodwill and dedication.

For some students, cheating is an option. Such a student may carry a crib into an examination, or copy assignments from friends. However, this raises a conundrum embodied in the question '*Can a true learner cheat on their learning?*' If a learner cheats, they cannot fulfil the objectives described above. They cannot achieve expansion of their own mind, because they are in effect cheating themselves. The concept of cheating becomes meaningless for a motivated learner. By cheating, you might get a result which pleases your teachers, which impresses your friends or which convinces your parents. In the end, however, there is no achievement, because *you cannot cheat yourself.* You know you cheated, and you know that you turned your back on an opportunity to develop your mind.

Intellectual honesty means being true to yourself. The internet offers opportunities to corrupt your intellect. Be aware of, and resist, the dangers. Life is short – don't waste your opportunities.

If these strictures are not sufficiently convincing then keep in mind the following. *An expert reader such as a university tutor can easily see the change in character of writing when a student has copied (plagiarised) something. Every writer has a style which is clear to a professional reader and after a period of time, the style of any particular student is usually quite recognisable.*

1.1.2 Assimilation

Research into learning strongly suggests that we integrate new learned material into the concepts we already have (Ausubel, Novak and Hanesian, 1978). In other words, the new knowledge joins in with the previous knowledge and out of this 'knowledge soup' new mental structures of knowledge develop. This constant process of re-integration of new knowledge with the old takes place throughout life and is called 'assimilation'. This concept will be returned to in subsequent pages.

1.1.3 Why Have We Gone Through a Discussion on Mind Expansion and Cheating?

As mentioned earlier, the internet is a repository of human knowledge which is expanding at a phenomenal rate. It is open to all who have the technology to gain access, and it will develop alongside, and interact with, human thought. Part of the associated technology allows us to copy and edit material we have obtained from the internet. We can electronically duplicate, cut and paste information to serve our needs.

However, if there is a need to satisfy an external approver (such as a teacher or an employer), it is tempting to copy and use material without assimilation. In other words, it is possible for you to copy, cut and paste so that the information has gone through your computer but *not through your mind*. Your mind has had no possibility to assimilate or expand, and you have cheated the external approver. The worst thing is that you have cheated yourself. As the technology gets better, and the ability to copy and cheat gets easier, the temptation to cheat gets stronger. Don't be tempted. If you cheat, you compromise your integrity. *Without integrity, you have nothing.*

Learning is difficult. The pain of learning is the hallmark of a stretched, adapting and developing mind. A tutor sometimes encounters a student who, worried about their performance, apologises by saying 'I've never done this before'. This worry is quite normal and an apology is unnecessary. The point of learning is to do things you have not done before . . . and it is certain to be uncomfortable. Like much in this life, if it is easy to get it is hardly worth having. Enjoy the challenge and try not to be tempted to take the easy way out.

1.1.4   Not Only the Internet!

This book does not propose that you study biology only through the internet, but that you use the internet as part of your *learning tool-kit*. There is a growing but misguided tendency to assume that the only information which is relevant and accessible is on the internet. Because it is easy to do, students increasingly search the internet for information via their computers at home or in work. Some students do this instead of visiting their local or university library. This can cause problems because in so doing, they might be using unreliable, unstable information.

Consider the following. Suppose I am taking a course in microbiology and have been given an assignment with the title 'Summarise the serological tests which might be used in the diagnosis of fungal disease in humans'. There could be three strategies.

Strategy one – use the internet

Using the keywords *serological+tests+fungal+disease*, a search engine yields 2010 hits. Most of the information on these 2010 web sites will be inaccessible, irrelevant or inaccurate. This could obviously take a lot of time, with no promise of success.

Strategy two – use a bibliographic search program

These are usually available via the university library. The same keywords as before are used via the Web of Science bibliographic search program and abstracts of the following articles are found:

● Del Negro G. *et al.* (2000) Evaluation of tests for antibody response in the follow-up of patients with acute and chronic forms of paracocci-dioidomycosis. *Journal of Medical Microbiology*, 49, 37–46.

- Derepentigny L. (1992) Serodiagnosis of candidiasis, aspergillosis, and cryptococcosis. *Clinical Infectious Diseases*, 14, S11–S22.

- Klingspor L. *et al.* (1996) Deep *Candida* infection in children receiving allogeneic bone marrow transplants: Incidence, risk factors and diagnosis. *Bone Marrow Transplantation*, 17, 1043–1049

- Nuorva K. *et al.* (1997) Pulmonary adiaspiromycosis in a two year old girl. *Journal of Clinical Pathology*, 50, 82–85.

- Swerdloff J. *et al.* (1993) Severe candidal infections in neutropenic patients. *Clinical Infectious Diseases,* 17, S457–S467

The problem here is that, in many cases, I cannot get access to the full articles on-line. I have to be sufficiently well organised to start my search early and then send for the reprints via the inter-library loan system.

Strategy three – use the library

The articles noted above might be available within the journals held by the library. While in the university library, I pick up a book such as the *Colour Atlas and Textbook of Diagnostic Microbiology* 5th ed. by Koneman *et al.* (1997) published by Lipincott-Raven and on p. 1058 find a table entitled 'Serological tests useful in the diagnosis of mycotic disease'. A sensible prose précis of this table could be useful.

The best strategy

The best strategy is, of course, to combine any of these approaches as effectively as possible. For example, a quick scan of the search engine results reveals a useful and reliable site at the University of Leeds in England at:

http://www.leeds.ac.uk/mbiology/ug/med/mycol.html#Candid_diagnosis 3/3/01–23/7/01

These are part of the support notes for the Microbiology component of the University of Leeds Laboratory and Scientific Medicine Course and there is some excellent information here. If this is *merged* with a review of the Web of Science articles, and a précis of the Koneman table in a sensible and integrated review, a positive outcome looks assured.

In recent discussions about web searching with academic colleagues at a number of international conferences, it has become evident that university tutors see web searching alone as a mediocre approach to doing assignments. However, the web is a good tool for *amplifying the value* of the material you have found through other methods of searching.

1.1.5   What to Print, What to Keep? A Problem of Using the Hardware

Almost everyone finds it more difficult to read large amounts of text on a screen than on paper. Consequently, most people print out the information they want on to paper, and then use it. What happens in this process ?

The source is visually, superficially scanned. If the material seems appropriate, large sections are printed on to expensive, high-quality paper, without assimilation of the information by the student. There are three aspects to the problem. You:

1 may not have achieved any learning,

2 may be called to account for plagiarism,

3 may cost the environment and your pocket with the amount of waste paper which is generated.

Try to cultivate the ability to scan information quickly on the screen, focusing on the essentials. Learn to use the tools. For example:

• *learn how to précis* – in other words to summarise a piece of someone else's writing in your own words. The easiest way to do this is to read a paragraph of the target text and then tell yourself what you think it means. As you are telling yourself, take down your own dictation. The précis will write itself.

• *learn how to use 'page preview'* in Netscape before you start wasting your paper, money and time. Incidentally, this facility doesn't exist in Internet Explorer and is the main attribute which makes the former system more useful than the latter.

- *download useful web pages* and keep them on file so that you can
  refer back to them, if you have enough space on your computer.
  This is especially useful if your tutor raises queries about the work
  you have quoted and the original web page has disappeared from the
  internet.

- *be careful about 'cut-and-paste' plagiarism*. If you want to cut and
  paste some text into an assignment, make sure that you put the text in
  quotation marks, and quote the source accurately using an appropriate
  system (see section 3.2). Be aware that excessive quotation can indicate
  intellectual sloth on the part of the writer.

1.1.6   Save Time and Money...

There are some tricks you can do which save you time and memory space
in looking at web pages:

- with Internet Explorer by turning off graphics,

  Tools>Internet Options>Advanced>Multimedia>then untick 'Show
  Pictures'

- with Netscape by making the print smaller,

  View>Decrease Font

- with Netscape by checking what you are printing before you print it,

  File>Print preview

1.1.7   Don't be Tempted to Use Cheat Sites

There are some sites which offer you the opportunity of buying
assignments which are already completed. Don't be tempted.

- Using a cheat site negates the whole reason for being at university.

- Unless they are custom written, they won't exactly satisfy the title of
  your assignments. One of the major faults in the normal student
  assignment is that the student did not address closely enough the
  question or issue that was posed.

- Your tutor will easily spot that the style isn't yours.

- Your tutor can also search these sites, using sophisticated electronic search methods, to find out where the text originally came from.

- Cheating ends up using more time and brain power than would be taken up by honest work – and the result is usually of a lower standard. Rely on your intellect; you have reached this level in the education system because you are actually quite good.

1.2   MANAGING FILES

Once you are able to search the internet, you will start to accumulate information that you wish to keep on file in your computer. These files will need to be organised in a systematic way so that you can retrieve data quickly on your own machine. Files and programs are stored in *directories*, now called 'folders' by Microsoft. When you take delivery of the computer, it usually comes with some file structure in place. For example:

- *executable files* (i.e. programs) can be stored in the folder called 'Program files'

- *data files* (i.e. files of text, images or spreadsheet data) can be stored in the folder called 'My documents'

You can create your own folders. Windows stores temporary swap files in virtual memory on the hard drive. As it fills up this short-term memory, it becomes slower and less efficient. You need to periodically log off, or switch off, or delete them directly to release these files and allow the computer to get up to speed again.

1.2.1   How the Computer Works with Files

In the computer, programs work on data according to your instructions, giving you a result (Fig. 1.1). The following examples show the *cyclical* way in which programs are used. Here 'data' can mean numbers or text.

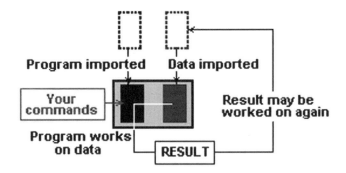

Figure 1.1   Iterative use of the computer

Example 1

You type text data into a document using a *word processor* and the document is amended according to your commands. You then put the document through the word processor and arrange the style of the presentation. You then put the article through a *spell check program* to correct the spelling.

Example 2

You have collected some experimental data. The data are fed into a *statistical package* which gives you a result. You then feed this result into a *graph-plotting program* which presents your results in a pictorial way.

Example 3

You feed data in the form of a keyword into a program which can search the internet. It gives you a result in the form of a hyperlink. You feed this hyperlink into a program which then displays the result for a web site.

These three examples show cyclical, *iterative* processes where you, the data and the program are working together in a highly productive way. However, you need some basic skills in order to select the appropriate programs and files. You get these skills through practice.

1.2.2 File Extensions

All the programs you will use are stored in *files*. Each file has a name. At the end of the name is a *file extension code* which tells the computer what type of file it has to deal with. Some of the most common files and extensions in Windows are as follows:

- Simple files of text have the extension .txt

- Text files with some, uncomplicated, formatting have the extension .rtf

- Files of text with sophisticated formatting have the file extension .doc

- Images have the file extension .tif or .gif or .jpg

- Sound files have the extension .mid or .wav or .mp3

- Excel spreadsheet files have the extension .xls

- Minitab statistics files have the extension .mtb

Try to work with files and extensions that are going to be readable on the computers you are likely to use. The usual problem is that you upgrade your computer to a higher level than that of your university (or they upgrade theirs). You then find that the lower-grade device cannot read the materials you prepared on the higher-grade device. Within this context, when deadlines are approaching, the coarse-sounding Sod's Law kicks in.

All practical scientists are familiar with this law which, summarised, states that *if something CAN go wrong, it WILL go wrong*. Thus the law says that:

> The computer on which text was written will always be of a much higher specification than the computer on which the text will be read. The text will then be unreadable on the lower-specification computer. The closer the deadline, the more likely this is to happen.

Incidentally, this law also says that *your printer will break down just before you are going to print out your assignment, just before the submission deadline runs out*. The answer is obvious. Leave more time!

If you are moving files of text from one computer to another, it can save a lot of trouble if you save the files as .rtf files before they are transferred, since .rtf files can be read by almost any word processor (Fig. 1.2). Apply a similar philosophy to all of your programs. For example, save Excel files into a low version of Excel before you move them from one computer to another.

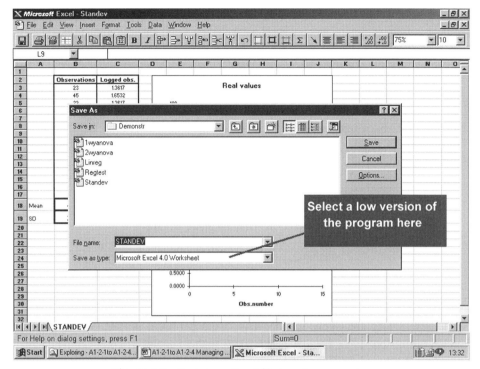

Figure 1.2 Saving an *Excel* file to a lower version

1.2.3 File Managing

The operating system of the computer offers you a way of moving, deleting and otherwise managing folders and files. In general terms, this is a *file manager*. In fact, there are two types of file manager. Early Windows systems (before Windows 3.5) had an excellent file manager (called, surprisingly enough, File Manager) which allowed you to:

- handle several directories at once,

- move things from one folder to another easily and smoothly,

- resurrect files that had been accidentally deleted,

- see the exact size of a file, thereby avoiding confusion between files with similar names (Fig. 1.3).

Although File Manager has been superseded by *Windows Explorer*

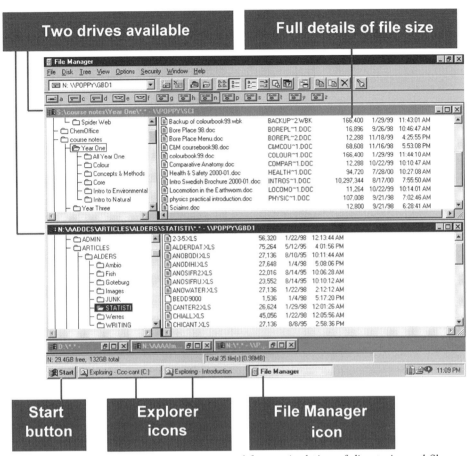

Figure 1.3   *File Manager* screen arranged for manipulation of directories and files

(Explorer), File Manager is still accessible in the Windows system up to Windows 98 2nd ed., and is a useful tool.

In later versions of Windows, in what appears to have been a process of reverse engineering, a file manager called 'Windows Explorer' was bolted onto the front of File Manager (Fig. 1.4).

However, you have to open two versions of Explorer to do what could be done with a single version of File Manager, and you cannot see the exact file size without extra mouse clicks, which wastes time and effort. In some Windows systems, file manager still exists and you can get access to it.

To do this, go to the bottom left of the screen, click on Start> Run and in the white box at bottom left type *Winfile*. File Manager will appear. However, you must be very careful. Although it is more convenient for file

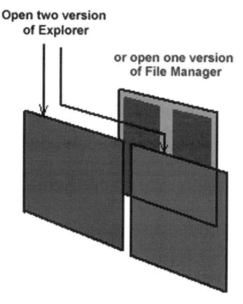

Figure 1.4   *Explorer* and *File Manager*

Figure 1.5   The *Explorer* icon

management than Explorer, *files deleted in File Manager cannot be undeleted. You have been warned*!

It is therefore safer, though less effective, to use Explorer. There is however, a trick which allows you to move files almost as effectively in Explorer as in File Manager. Click on the Explorer icon (Fig. 1.5) twice so that you have two versions of Explorer open. If you then put the cursor on a piece of grey bar at the bottom of the screen and click the right mouse button, you can arrange both Explorers on your screen at once (Fig. 1.6). It is then easy to move files from one Explorer file structure to the other. For example, in Fig. 1.7, a file called Doc3 is being moved from one part of the

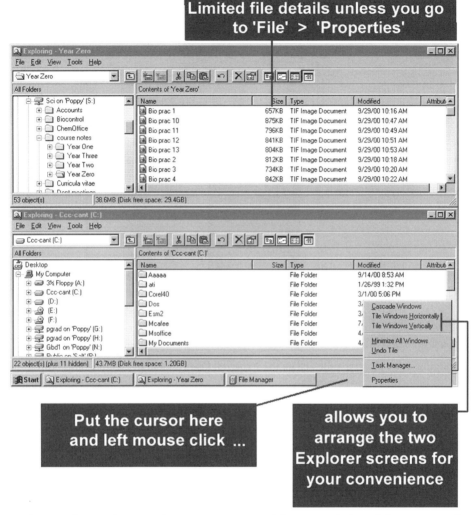

Figure 1.6 Windows Explorer screens arranged for manipulation of directories and files

Explorer to another. This is achieved by highlighting and dragging it from a location in one of the Explorer windows on the left and dropping it in a destination in the Explorer window on the right (drag and drop).

Why use File Manager at all? Periodically things go wrong with Explorer and you can trace the faults by use of File Manager. Also, if you are preparing (say) a thesis or dissertation, with many different versions of chapters, it is valuable to be able to see the exact file sizes. This is easier in

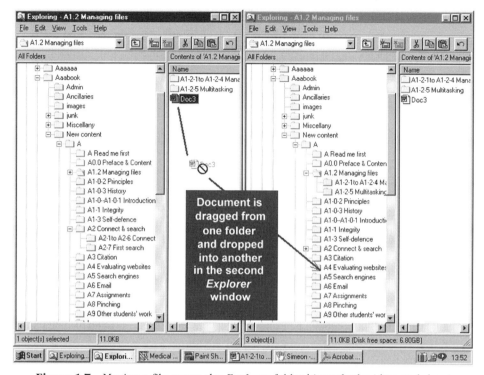

Figure 1.7   Moving a file across the *Explorer* folder hierarchy by 'drag and drop'

Folders   Files

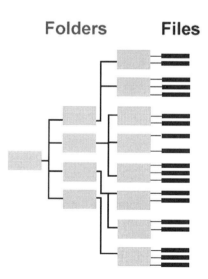

Figure 1.8   Hierarchical arrangement for keeping tidy folders

File Manager than in Explorer. However, remember never to delete files. *Always put them in a 'junk' folder.*

1.2.4 Tips on File Handling

Once you are adept with a file managing system, you will need to arrange your files in some sensible order. Here are some tips.

- Organise your files in a branching *hierarchy* (Fig. 1.8).

- If a range of files have been accumulated which relate to particular web-page files or .doc files, it is useful to keep each set of files together in a separate folder.

- Create a *'junk folder'* in which you can put earlier versions of a piece of work. Never delete files in the junk folder until you are absolutely sure that you have finished the work.

- Prepare a large document such as a dissertation or long essay as several small documents of one or two pages. This will save time in opening and closing documents. There is also less danger of accidentally losing the whole thing.

- If there is a folder which you use frequently, begin the folder name with AAA. It will then be placed at the top of the file structure and you won't have to scroll down to find it.

- Frequently, you will save files into the wrong folder. Become adept at using Windows Explorer>Tools>Find> to locate misplaced documents. In File Manager it is File>Search.

- Use a file or folder beginning with AAA to sit right on the top of your file structure. Use this for saving files which come attached to email messages. You can then deal with them at leisure, dragging them away to where they belong or deleting as necessary (see section 6.2).

If you misplaced the document in Word and worked on it quite recently, use the 'File' drop down menu. At the bottom of the menu are displayed the last four files you used, numbered one to four. Summon the file of interest by clicking on its name. When the file appears, immediately go to File>Save As> and find out where the file has gone to. You can then use Explorer or File Manager to drag it to its proper location.

Schofield (2000) suggests that if you divide the number of directories by

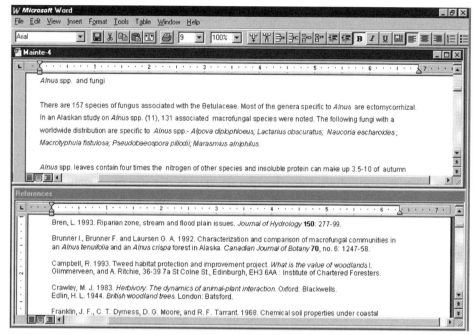

Figure 1.9 Two different word files, opened on the same screen. In this example, references in the bibiography can be cross-referenced to the main text

the number of files you have, you can get a measure of your organisational efficiency. If you have more than 500 files in a folder, neither you nor your machine can work efficiently.

1.2.5 Multitasking

It is possible to use several programs at once. This is called 'multitasking'. You could word-process a document at the same time as you are searching the internet. Since it takes some time for a search to end, multitasking means you can do something useful instead of merely watching the screen blinking at you. Here are two examples.

Example 1

When preparing a complex document such as a dissertation, the text may be in one file and the references in another. It is convenient when checking references to have both Word files open at once. If you are working with a

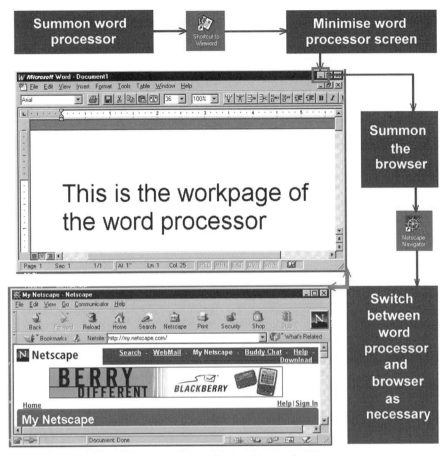

Figure 1.10 Sequence of actions for working with two programs at once

Word document and images, it might be useful to have the word processor and the image processor open at the same time (Fig. 1.9).

Example 2

Let us presume for the moment that you can use a *web browser* such as Netscape. The word processor Word (Fig. 1.10) is summoned by a double click on the appropriate icon on the *desktop* (the main screen you see when you switch on the computer). Minimise the Word screen using the 'negative sign' icon at the top right of the screen. The Word icon then appears at the bottom of your screen to show you it is still running. Place the cursor over the browser icon, in this case Netscape, and double-click

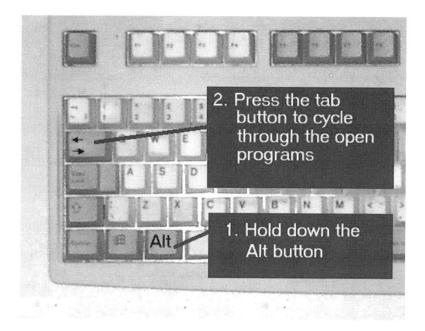

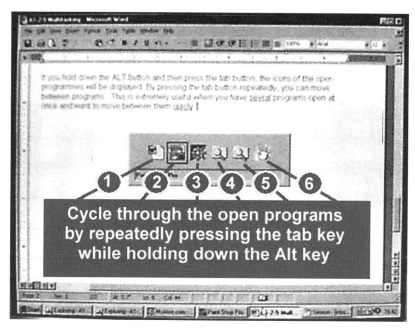

Figure 1.11   Cycling through open programs using the Alt and tab keys

the left mouse button. The browser screen appears. This too can be minimised. You can then switch between Netscape and Word depending on what you are doing.

If you hold down the ALT button and then press the tab button, the icons of the open programs (sometimes called 'applications') will be displayed. By pressing the tab button repeatedly, you can move between programs. This is extremely useful when you have several programs open at once and want to move between them quickly (Fig. 1.11).

1.3 SELF-DEFENCE

1.3.1 Introduction

There are routine precautions which everyone should take when using the internet. Use your integrity and common sense. For example, only reveal your name and address to people you can trust absolutely, and remember such people are few. Certainly don't reveal this information on your personal web page. If you have a child in your care who has access to the web, then make sure you know your responsibilities. Valuable guidelines are offered by Lawrence Magid, a columnist for *the Los Angeles Times* at:

http://www.safekids.com/child_safety.htm 3/4/01–25/7/01

In an academic context, you need to defend yourself against two threats on the internet:

1 unreliable, irrelevant or invalid information;

2 viruses, both malicious and innocuous.

Unreliable information can be that posted by students on the internet who have less understanding than you do. *Irrelevancy* is a risk which arises when you don't assimilate material. If you download chunks of material without reading it and then cut and paste it into a document, your tutor may find it irrelevant. It could also be plagiarism, of which more later. *Invalid information* is information which does not stand critical scrutiny.

1.3.2 The Need to be Critical to Identify the Invalid

If you are going to use biological information from the internet, you need to be clear about *how you will use* the information you have acquired. Most

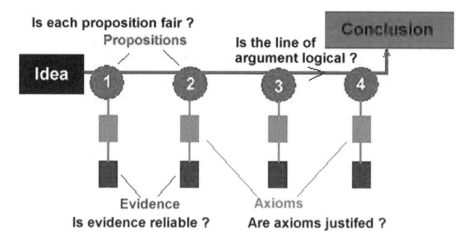

Figure 1.12   Map of an argument

academic writing, in subjects as diverse as biosciences and history, involve a process of *argument* (Fig. 1.12). Evidence is used to crystallise facts which are used to develop fundamental truths called 'axioms'. From an axiom comes a *proposition* about the evidence and the axiom. Propositions may be linked together to make a *line of argument*. The line of argument should lead to a *conclusion* which is *justified* by being, to a lesser or greater degree, logically dependent on the propositions.

For example, suppose I want to make an argument concerning evolution. There are two definitions of evolution:

1 *Evolution is a change in gene frequency through time.*

2 *Evolution is a change in functional morphology through time.*

I will use the second definition.

Idea   Functional morphology changes through time.

Proposition 1   Poorly adapted organisms are less fit.

Axiom   There is death and survival in any cohort.

Evidence   The fact that I am alive and other humans with congenital abnormalities such as microcephaly are dead.

Proposition 2   A parent produces offspring which resemble the parent.

Axiom	There is inheritance of genes.
Evidence	In terms of functional morphology, I resemble my parents (one head, two arms, two legs, vertebral column, etc.).
Proposition 3	A parent produces offspring which vary from each other.
Axiom	Meiosis and sexual fertilisation will produce variety in a cohort.
Evidence	I do not have exactly the same morphology as my brothers and sisters.
Proposition 4	Environmental change operates on all of the above.
Axiom	Changing environments will cause death of some organisms.
Evidence	I was involved in an accident in which people who could not swim died. I have the functional morphology which allows me to swim, and I survived (apocryphal – I'm only making a point).
Conclusion	If all of the above propositions are accepted then functional morphology of humans must have changed through time.

Why is this conclusion justified? In changing environments (for example, during the extension of pre-human *Homo* across the world, those *Homo* who were functionally adapted survived and passed on the genes for their success to their offspring. Those who were not died out. As this happened over successive generations, and as the environment changed around them, functional morphology changed accordingly. For example, Inuit (North American Eskimo) do not have the same functional morphology as Masai cattle herders in Africa.

You can make an argument as:

- part of a practical write-up,
- as an essay,
- as part of the answer to a quantitative problem,
- during an examination,

- in the union bar, where your proposition is that one football club is better than another,

- with your partner over whether to buy this or that sofa.

In each case, the quality of your argument will be judged on the basis of:

- general validity and relevance of evidence,

- justification of the axioms based on their evidence,

- justification of the proposition arising from their axioms,

- whether the propositions are logically linked,

- whether the overall conclusion is justified.

Just as a tutor will judge your work on these criteria, you should use the same *critical judgement* on what you encounter on the internet. What follows is a true story about a situation in which a student did *not* use their critical faculties.

A group of students were set an assignment concerning *Biological Aspects of Global Change* and asked to prepare a dossier based on information drawn from the internet. One student in particular had produced an apparently excellent assignment as a hypertext-based series of web pages with links to other web sites. However, careful reading of the assignment suggested that the information had not been sufficiently critically evaluated, and the tutors decided to investigate further by means of a tutorial, with the student, at the computer. Here, the student was asked to explain the layout and development of the assignment and was also asked to explain aspects of certain crucial articles which had been cited.

The student had found a well-written article on world population that described basic aspects of human demography. In this article, the argument developed rather neatly into an apparently innocuous statement that many of the problems of world population could be traced to an '*excess of governmental control*'. The significance of this statement varies depending on both the context of the reader and the context of the writer.

- *The context of the reader*

  To a politically right-wing reader, the statement might seem perfectly acceptable; any kind of state service including an education or health system might constitute excessive government. A politically left-wing reader might challenge the idea of excessive government. Who determines the threshold over which there is an excess? The extreme monetarist of the Thatcher years in the 1980s might read excessive government differently from the centrally planning socialist in the era of the Wilson government of the UK in the 1960s.

- *The context of the writer*

    In the tutorial, the hypertext links were followed in order to find out more information about the writer. By consulting his home page, the author of the article was found to be a consultant for a major international corporation making disposable paper-based products. A cynic might say that his view was driven by the opinion that governmentally controlled regulation (for example of child labour in developing countries) in turn might lead to lower profits.

The point at issue is not that the writer and reader are adopting particular stances but that the student had not read the articles sufficiently critically to appreciate that (i) there might be a range of views about the subject, and even more importantly (ii) the author *might* have a hidden agenda.

1.3.3 Malicious Viruses

Viruses are malicious programs which cause inconvenience or damage to innocent people. The 'Lovebug' virus originated in the Philippines and infected 6 million computers in May 2000. When people opened the email message 'I love you', the virus copied itself, obliterating other files and then sending itself as an attachment to everyone in the user's email address book. It caused more than 1000 million dollars' worth of damage. During the Cold War, people were employed on both sides of the Iron Curtain to produce viruses that could be aimed at the computers of the opposing side.

Most commonly, viruses are received when a file attached to an email message is opened. It is a good idea to not open such attachments unless you are sure of the provenance of the file and are certain that it is innocuous.

1.3.4 Innocuous 'Viruses'

It is difficult for the non-*cognoscenti* to differentiate these from real viruses. The best solution is to ask a member of the *cognoscenti*, usually a computer buff or your helpdesk. You can find out about such hoaxes by visiting the US Government Incident Bureau at:

http://www.ciac.org/ciac 5/7/00–25/7/01

There are three major types of innocuous viruses – hoax virus messages, heartstring pullers and jokes.

- *Hoax virus messages* – these appear periodically by email and are passed on by concerned colleagues and friends. Frequently, they are claimed to have been presented as an alert on a particular news service. A typical example was a hoax virus called 'Good Times', which was circulating in May 1995.

- *Heartstring pullers* usually come in the form of a hoax message about a child who is dying of some dread disease and wants as many people as possible to do something like send an email message to an address, and then pass a request to all acquaintances to do similarly. If everyone spends only one minute on this hoax, an enormous amount of human time is wasted.

- *Jokes* are a kind of innocuous 'virus'. They flash across the internet and most regular internet users will receive at least one joke per day from friends and colleagues. It is a good idea to have a special directory or folder in your email system to which you can save these jokes. They do not all have to be looked at when they arrive, if at all. Like many colleagues, I now delete any joke which has an attached image since I don't have the time to store it and open it.

1.3.5 Precaution Checklist

- Use common sense.
- You cannot be too cautious,
- But don't be paranoid.
- *Never* give out your personal phone number or address.
- Use a pseudonym; if your real name is Josephine Bloggs, give yourself a

false name when talking to people on the internet who don't know you in the flesh.

- Sites which want personal information want it for a reason. They may sell it on, and then you will receive *spam*. Spam is the equivalent of paper junk mail. It usually tries to sell you something you don't want. You will sometimes receive spam which will ask you to send an email disconnecting you from that particular source. Often, the perpetrator uses these bounce-back emails to confirm that you are still looking at their spam. The best thing to do is just to delete spam messages.

- Do not open any attachment whose source you don't know and trust.

- Periodically check up on virus alerts via

    http://www.virusbtn.com 5/7/00–25/7/01

    http://www.wildlist.org 5/7/00–25/7/01

  You can buy protective software via

    http://www.symantec.com 5/7/00–25/7/01

    http://www.sophos.com 5/7/00–25/7/01

    http://www.mcafee.com 5/7/00–25/7/01

If you work on a network which is controlled by an institution such as a university, there will almost certainly be a *firewall*. This is software which protects the system from outside problems such as viruses or hackers. You should keep in mind that if a network has a firewall, it is almost certainly monitored in one way or another, so don't do naughty things such as libelling the vice-chancellor, downloading pornography or trying to hack into the Pentagon from a network-controlled computer.

Finally, consider this virus warning which appeared on the web:

An email from Heaven

I was given this message. This virus is very nasty. Pass it on to all of our friends.

VIRUS WARNING
If you receive an email entitled <Badtimes>, delete it immediately. Do not open it. Apparently this one is really nasty. It will not only erase everything on your hard drive but it will also delete anything on disks

within 10 metres of your computer. It demagnetises the strips on ALL of your credit cards. It reprograms your PIN access code, screws up the tracking on your video and uses subspace field harmonics to scratch any CDs you attempt to play.

It will recalibrate your refrigerator settings so all your ice cream melts and your milk curdles. It will program your phone Autodial to call only 0898 sex line numbers. This virus will mix antifreeze into your fish tank. It will drink all your beer. It will leave dirty socks on your coffee table when you are expecting company.

It will replace your shampoo with engine oil and your engine oil with orange juice, all the while dating your current girl/boy friend behind your back and billing their hotel rendezvous to your VISA card. It will cause you to run with scissors and throw things in a way that is only fun until someone loses an eye.

It will rewrite your backup files, changing all the active verbs into passive tenses and incorporating undetectable misspellings which grossly change the interpretations of key sentences.

If <Badtimes> is opened in Windows 95/98, it will leave the toilet seat up and your hair dryer plugged in dangerously close to a full bath. It will also molecularly rearrange your aftershave/perfume, causing it to smell like dill pickles.

It will install itself into your cistern and lie in wait until someone like your new girl friend, does a serious defecation, then block the U bend and cause the toilet to overflow.

In the worst case scenario, it may stick pins into your eyes.

PLEASE FORWARD THIS WARNING TO EVERYONE YOU KNOW.

2
Getting Access to the Internet

2.1 INTRODUCTION

No matter how you connect to an intranet or internet, you will have to use a *username* and a *password*.

• The username will usually involve some aspect of your real name, for example, your initials.

• The password is a *unique code* that only you will know. Choose something reliable and easy to remember. You are always cautioned not to write it down. No one pays attention to this, and everyone always writes down their password; make sure that it cannot be found by anyone other than you, accidentally or on purpose. If security is a major concern, don't use a word such as 'daisy' or 'biology'. Hackers can create programs which will automatically cycle through a dictionary of words until they find the correct one. Use a random arrangement of numbers and letters.

The first time you connect to the intranet (called 'logging on') you will have to put in the username, the password and then repeat the password to confirm it. Logging on derives from the fact that the activity of every user is monitored ('logged') by the computer in the sense that it 'knows' you are working on-line. The first word to be used on the internet was 'log' (Naughton, 1999).

2.2 CONNECTING UP VIA AN INTRANET

The best way of being connected to the web is to get someone else to do it and pay for it. If you are in an academic institution in the UK, it is probable that your institution is connected to the *Joint Academic Network (JANET)*. Most institutions have an *intranet*. This is an in-house net comprising all the linked computers in the institution protected from the main internet by a *firewall*. The university will have a computer services department which maintains and administers the computers and a *Helpdesk* to give immediate software and hardware advice. You will have to approach your computer manager or Helpdesk who will issue you with a username and password.

Most educational institutions have *open access areas* (OAA) where computers are generally available. However, they are usually fully occupied during the day and there may be a booking rota. If you are an average student, think about the times of day when it is inconvenient for you to go the OAA (e.g. meal times). At these times it is probably just as inconvenient for others, so the OAA will probably be quiet. Arrange to go at these times. Chellen (2000) provides a comprehensive account of the procedure for getting access to a university network.

Incidentally, be aware that your transactions on the web via JANET may be more or less stringently monitored by the computer services department. Unsavoury web sites are automatically blocked. If you are discovered to have been using JANET for illicit purposes such as downloading pornography, you may be in serious trouble.

Most universities limit the amount of free printing you can do. Be selective about how many web pages you print as paper. Although it is hard work, try to read material on the screen rather than printing it out. That way, you will assimilate lots of valuable material without necessarily printing it. Only print what is absolutely essential. If there are images you don't need, some systems allow you to disable the image facility so that you are only printing text. Alternatively you can save the web page and look at it at your leisure.

If you are not linked to an intranet, probably working at home, you will have to get connected to the internet. As houses accumulate computers, it is increasingly probable that each home will have its own intranet. For example, several computers might be linked to a single peripheral such as a scanner.

2.3   CONNECTING UP INDEPENDENTLY

See Fig. 2.1. To get on the internet, you need:

- a *computer*;
- a *monitor*;

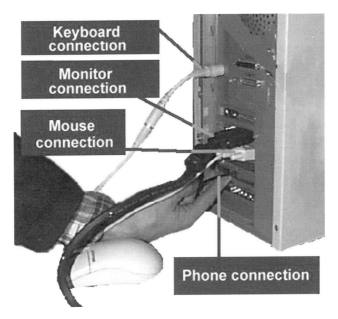

Figure 2.1 The modem link through the telephone line

- appropriate software on the computer – for example, an *operating system*; the most common system in current use is Windows but there are others such as Linux;

- a *modem* – this used to be a separate box but nowadays comprises a modem card in the computer. A minimum specification is V.90, which receives at 56 kilobytes per second;

- a *telephone line* connecting the phone network to the modem;

- an *internet service provider* (ISP).

The internet works through a *transmission control protocol* (TCP) which splits messages into packets and routes them. High-speed data lines called 'backbones' connect the major parts of the internet.

When you buy a computer, it comes with comprehensive but simple instructions for connecting everything together and then getting on-line. Many of the *devices* such as scanner and printer will need to have *drivers*. These are programs (software) which are installed via the CD drive. This, too, is quite simple. Once you have put the appropriate CD in the driver, the software will often load itself. If you have problems, try to consult friends rather than the helplines. The latter are usually expensive, currently 50p per minute.

Data are (note that 'data' is a plural noun – the singular of data is 'datum') passed over a telephone line as *DSL – digital subscriber line transmission.* *ISDN* (*internet subscriber digital network*) transmits at 65 kb/s and a faster alternative is *ADSL* (*asynchronous digital subscriber line*) which transmits at 1024 kb/s.

2.4   *CONNECTING UP VIA AN INTERNET SERVICE PROVIDER (ISP)*

If you are not on an intranet, you will have to make your own arrangements for connecting to the internet. You do this through an internet service provider. Typical ISPs include BT, AOL, MSN and Freeserve. Each ISP offers a different range of services. Research pays off. ISPs were reviewed in the magazine *Internet Advisor* in July 2000 and there are some ISPs which offer a free service, making their money by advertising. There is a trade organisation for ISPs at:

http://www.ispa.org.uk 8/10/00–23/7/01

You contact the ISP by letter, telephone or email (usually the latter) and they negotiate a username and password with you.

2.5   *BROWSERS AND GENERAL ASPECTS OF SEARCHING*

Browsers such as Netscape, Internet Explorer or Opera are usually available free. You get them either when you buy the computer or by downloading from the internet. The browser will install itself on your computer.

At your computer, you load up the browser and connect to the internet by double-clicking on the icon (Fig. 2.2). There are two ways of getting to your target subjects:

- You can type, or preferably copy, a *web address* (*URL*) into the *location bar* (see section 2.6) at the top of the browser screen. When you hit the 'return' button, you will be taken to the target page.

- You can do a *search*. Most advanced browsers and web sites now offer a search option, in which you type a target subject (in this case it is the

Figure 2.2 Browser icons

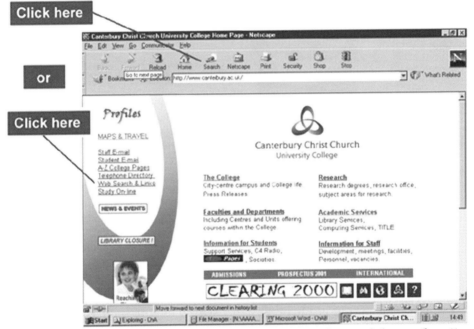

Figure 2.3 The Netscape Communicator page, showing the search button (http://www.cant.ac.uk)

subject of interest, not a URL) into a search box. The browser then supplies a list of what it considers to be sites relevant to your interest. These sites are *hyperlinked*; in other words, when you select one, you go direct to its web site. Part 2 of this book takes you through a set of example searches.

The examples in Figs 2.3 and 2.4 show the front pages of two common browsers. Both are looking at the same university *web page*. The university

Figure 2.4   The Microsoft Internet Explorer page, showing the search aspects (http://www.cant.ac.uk)

web page is shown on the bottom-right half of the screen. It is clear that the arrangement of the browser page around the top and left-hand side of the university web page are similar.

The browsers usually have a word or button which show that you can do a search. This button is a hyperlink to a *search engine* such as Google. In the associated box, you type the title of the thing you are looking for, press the search button and the search engine looks for any relevant web pages. The search engine lists these subjects, together with a little extra information from each web page. Each entry in the list has its own hyperlink which means you can go directly to the site which interests you. Note that:

- searching in lower-case will bring in both lower- and upper-case information, so don't use the shift key;

- if you connect words with a plus sign, the engine will look for pages which include *all* the words you have suggested, *in any order*.

If you want to search by a phrase – for example, for information on human evolution, then place your phrase *within double quotation marks*, e.g. "human evolution". The engine will search for an exact match to this phrase.

Each search engine has its own protocols. Some are rather primitive; for example, they will give you sites to do with *all* of the words you have put in the search box, ignoring your logical terms. For example, when I used the search terms *pig AND iron* or "*pig iron*" to find something about iron smelting, with either method, UKPLus gave 332 references, mostly to pig farming. Find a favourite engine and get familiar with how it works. The current favourite of most academics is Google (section 5.3).

2.6   THE FUNDAMENTALS OF A WEB SITE

The user writes a web address beginning with http:// in the long white *address box* at the top of the page, sometimes called the 'location bar'. On pressing the 'return' key, the computer then starts a hunt for the web site which is held at this web address. Pages from the web site are then transferred (*downloaded*) into the user's computer. The user then clicks hyperlinks on these pages to go to other pages on this site or other sites. A followable link is usually indicated when the cursor arrow is replaced by a *pointing hand* (Fig. 2.5). It is an invitation to follow a link.

The *scroll bar* on the RHS is used to move up or down the page. The *print icon* is clicked to print. Pages can be *bookmarked*, copied and saved. Bookmarking is called 'saving to favourites' in Internet Explorer.

Web sites are collections of web pages, written in *html language* (see Introduction). The pages are constructed on a user's computer and then uploaded to the internet service provider (ISP) server by a *file transfer protocol* (FTP). This is usually done relatively automatically via software controlled by the ISP. Other users can then get to look at these web pages when they are allowed to do so by the ISP.

Web sites can be constructed via a number of programs including Microsoft Publisher, Serif PagePlus, Microsoft Frontpage, Netscape Communicator (Composer) and Dreamweaver.

2.7   BOOKMARKS AND FAVORITES

Once you have found an appropriate web site, you may want to return to it in the future. *Bookmarking* is a method of preserving the web addresses (URLs)

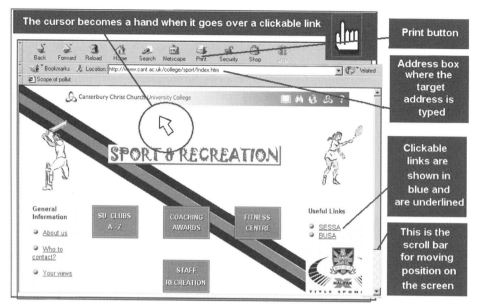

Figure 2.5 Features of a typical web page (http://www.cant.ac.uk/college/sport/ Index.htm 2/6/99–23/7/01)

of sites which you have found to be useful. Let's presume you are concerned about health and safety in the laboratory or in the field. Using Netscape you have found an interesting site on first aid, offered by the BBC at:

http://www.bbc.co.uk/health/first_aid_action/index.shtml 5/3/01–26/7/01

Use the mouse to select the Bookmark button, which gives you a grey *drop-down menu*, listing your bookmarked websites (Fig. 2.6) At the top of this list, you have the option of adding, filing or editing bookmarks. Select 'Add bookmark', and the web site is now added to your list of bookmarked web sites. To view the list in more detail, select 'Edit bookmarks'. If you then use the mouse to double-click on any of these bookmarks, you will be taken directly to the web site. (This isn't quite true. It *feels* like this, but what actually happens is that web pages are downloaded to your computer.)

In the Internet Explorer system, the favorites end up in a screen behind the main screen and you have to drag the main screen to one side to see them. However, you can temporarily remove the favorites by a click on the X at the top right of the favorites screen and return to it by re-clicking on the favorites button. The Internet Explorer system seems to be more clumsy than Netscape in this regard.

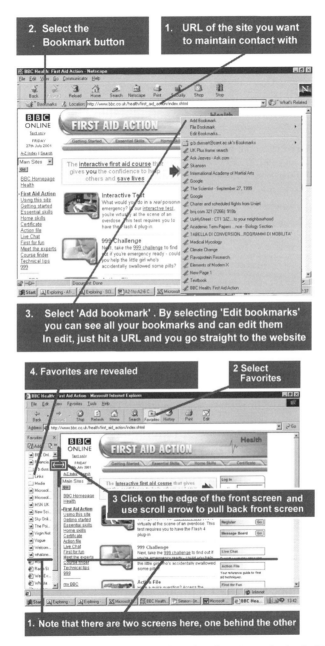

Figure 2.6 Steps to take in order to bookmark a favorite web site in Netscape and Internet Explorer. (Reproduced with kind permission of BBC Health at www.bbc.co.uk/ health/)

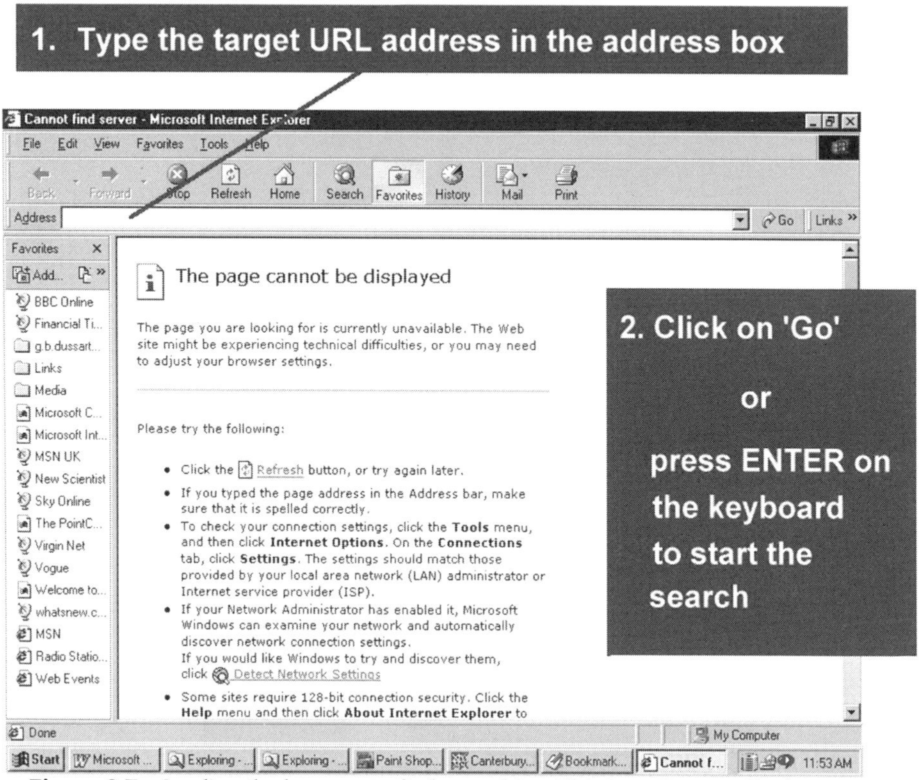

Figure 2.7   Sending the browser to find something. This is an *Internet Explorer* page. *Netscape* is similar

2.8   YOUR FIRST SEARCH

2.8.1   Making a Search – the Client Requests...

After you have connected up and switched everything on, you can do your first search. Click on the icon for your browser (see section 2.5). The opening screen of the browser will be downloaded to your computer. Type the target web address (the URL) in the address box at the top of the screen (see Step 1, Fig. 2.7). Now follow this example where I am going to have a look at the weather forecast. I type in:

http://www.meto.gov.uk/sec3/sec3.html 3/7/00–26/7/01

The cursor is moved over the 'Go' icon, or the 'enter' (sometimes called 'return') key on the keyboard (see Fig. 2.8). The computer tells the ISP server that it wants web pages for this address.

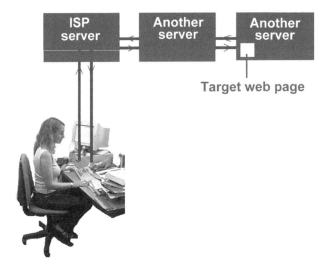

Figure 2.8 Your query and response travel via the servers

2.8.2 ... and the Server Responds

The browser searches the web (or searches its own lists) until it gets to the server which is holding the target web page (Fig. 2.8).

The target web page is then *downloaded* to your computer. In this case, until this web page is changed,* the screen looks like that in Fig. 2.9.

2.8.3 Avoiding Distractions

Try to *keep focused* when searching. Most web sites make their living by selling advertising space, so they are bound to be distracting. Very capable and highly paid people are spending their lives trying to distract you into looking at their advertisements on web pages. The many distractions will include advertisements, promises, titillations, music files and glimpses of a host of subjects you would really like to follow up. The Discovery channel has some interesting scientific news material. Its web site is at:

http://www.discovery.com 4/4/01–26/7/01

---

*Although many web sites recommended here will stay on-line for a considerable period, during the life of this book some web pages will certainly change. It cannot be guaranteed that you will see the web pages exactly as they are shown here, but the principles endure.

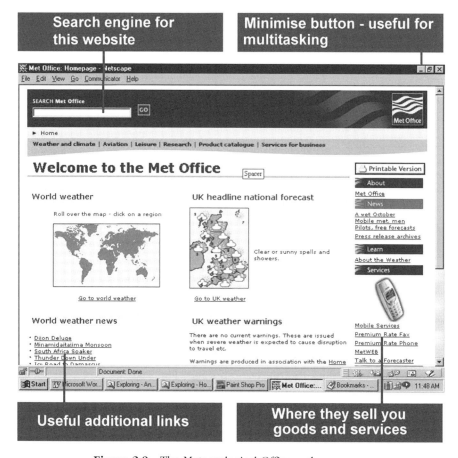

Figure 2.9   The Meteorological Office welcome page

However, the web page typically tempts you into distraction, for example, window shopping or crosswords. One solution to distractions is to *Bookmark* the site and return to it when you are more focused.

Become adept at using Bookmarks in Netscape or Favorites in Internet Explorer so that you can look at these web pages at your leisure (Fig. 2.10). Above all, stay focused. Remember that you have a deadline and that a lot of this additional material has actually been constructed to distract people just like you. Set yourself a pre-deadline deadline. Submit your assignment with days to spare and then you can spend time looking at these other areas. Keep your bookmarks tidy by appropriate housekeeping (edit the file by cutting and pasting) and also periodically *save back-up copies* of the bookmark or Favorites file to a floppy disc or onto your hard drive.

Figure 2.10 EnviroZine. A Canadian on-line environmental magazine which could be worth bookmarking. http://www.ec.gc.ca/EnviroZine/english/about_e.cfm

To return to your bookmarked pages in Netscape, select 'Bookmarks' and then select Edit Bookmarks. A list of your bookmarks will be presented to you and a double click on any one will take you straight to the relevant web page. A similar procedure is used for Favorites in Internet Explorer (see Fig. 2.6).

2.8.4 What is the Composition of the Web Address URL (uniform resource locator)?

Knowing how a URL is composed can be useful. It gives you clues as to the nature of the web site and even the author. The URL is made up of several parts.

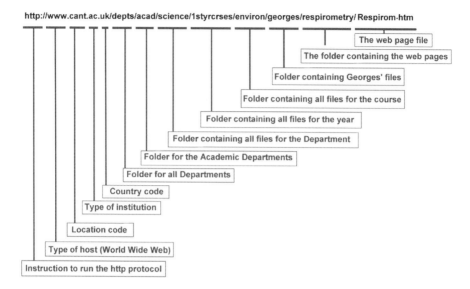

Figure 2.11   Analysis of the URL of a typical university web page

The first part describes the *type of information* on the web page (http, ftp, etc.). It is a code for the type of access involved, e.g. http:// ftp:// gopher://

http stands for *hypertext transfer protocol* and specifies that there will be a

1 connection

2 request

3 response

4 closure

The second section *locates the server*. Later sections *navigate the search engine* through the file structure of the host server until the web page is found (Fig. 2.11). After this come the *domain names*.

If you are transcribing a URL by typing it out, each part of the URL must be typed accurately, observing spaces and cases. Be especially careful with *underscores* (_) and *tildes* (~). If possible, it is best to cut and paste URLs into the search bar (Section 3.2). Make sure that at the end of the URL, after typing .com or .uk or .html, you do not insert an *extraneous full stop*. For example:

http://cant.ac.uk *will work*

http://cant.ac.uk. *will not work*

2.8.5 Types of Organisation which can be Identified from the URL

.ac   academic institution such as a university or college in the UK

.edu   academic institution such as a university or college in the USA

.com   US company

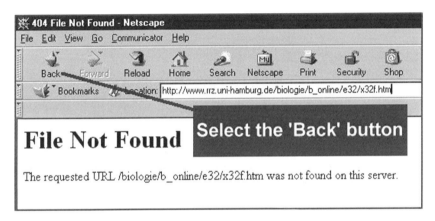

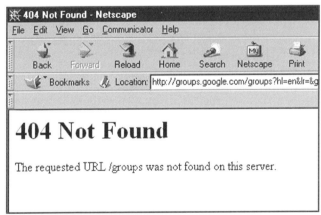

Figure 2.12 'Not found' error messages. This is a dead end, so go back via the BACK button

.co   non-US company

.gov   UK government department

.net   internet service provider (ISP), e.g. AOL or Freeserve

.org   non-profit organisation

.sch   school in the UK

2.8.6   Error Messages

The code for a web page which cannot be found is ERROR 404 (Fig. 2.12).
If you get this, hit the BACK button in the browser until you reach a
working page.

3
Citation Guidelines for the Use of Internet Materials by Students

3.1 INTRODUCTION

In non-fiction writing, any reader should be able to follow an *audit trail* to the source of the information. The reader needs to be able to verify the truth of what is being stated. Everything should be justifiable. This is a watchword in academic writing, whether by a professional writer or by a student.

If you are a student, try to put yourself in the shoes of your tutor. One of the major reasons they do the job at all is because they like to learn too. They are not reading your work merely to give you a mark. Tutors are interested in your ideas and in the material that you put into assignments and will often be intrigued enough to follow up your references.

Any reader or examiner will judge the relationship between the student's own words and the internet links by the criterion of *fitness for purpose*. In other words, web-based materials or electronic materials should not be too extensive for the reader. Be critical about the quality and quantity of the material you are invoking from the internet.

If you quote a filename, always *give it in full, including extensions*, e.g. .doc .jpg .zip. Preferably *paste* the URL into your documents. If a tutor spends time trying to follow a URL you have cited, you will get black marks if the URL turns out to have been incorrectly cited. Tutors are jealous of their time and don't like to have it wasted by carelessness.

3.2 ADVICE ON CITING AND LISTING URLS

When you have found a web site which you want to cite, it is important to *cite the URL (the web address) very accurately*. In the past, when the

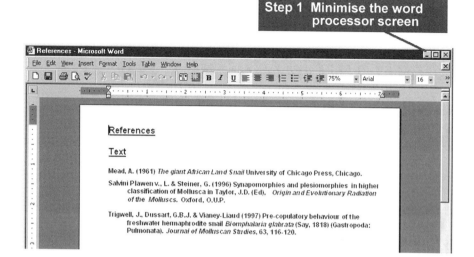

Figure 3.1   Starting the reference list in the word processor

tutor had to leave the office and go to the library to follow up a reference, you could have gambled with the accuracy of your citation in an assignment. Now, you should remember that your tutor will have easy access to a browser and will almost certainly investigate the URLs you have quoted. Remember that even a full stop out of place will make the URL inaccessible.

For reasons of accuracy, learn to cut and paste URLs by following these steps:

1   Have your assignment ready on the word processor page and then minimise the screen (Fig. 3.1).

2   Contact your ISP and use your browser to target the web page you want.

3   At your target web page, *single-click* somewhere on the URL in the address box. This will highlight the entire URL (Fig. 3.2).

4   Use the 'copy' command to copy the entire URL into the reference list.

5   Minimise the web page and restore the word processor page.

6   Use the paste command to place the URL where you want it in the assignment (Fig. 3.3).

7   Restore the web page.

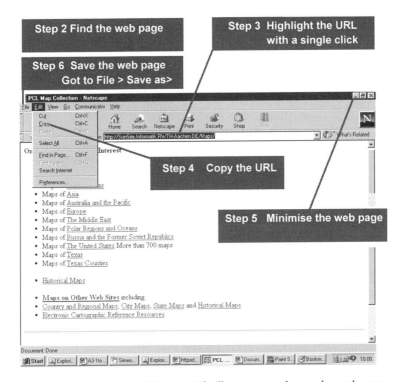

Figure 3.2 Copy the URL to avoid silly errors, and save the web page

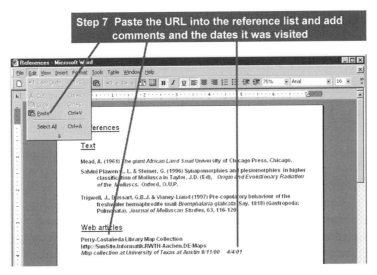

Figure 3.3 Pasting the URL into the reference list

8 Save the web page into your files for further reference. You can do this by selecting FILE in the command line at the top of the screen. A drop-down menu appears. Select 'SAVE AS' and then put the web page into an appropriate folder.

9 If the web site is really worthwhile, Bookmark it (see section 2.7).

3.3 TEN COMMANDMENTS OF CITATION

The following provisional guidelines are offered as a way of improving the quality of student use of the internet. Although the guidelines are directed towards the work of bioscience students, they should be applicable to anyone using electronic media as a source of information.

1 Students should *critically evaluate* all material.

2 The URL should be quoted *in situ*, so that the context is clear, and *also* should be presented in a reference list at the end of the work. This URL list should be separate from a reference list that relates to non-web documents such as books and articles.

3 The reference at the end should include:

- the *name of the authors*, if you can identify them
- the *title* of the work (you may have to write this yourself)
- the *full URL address*
- *explicit dates* when you visited the web site concerned
- a *justification*.

4 To reinforce the last point ... the use of a particular web site should be *intellectually justified in the reference list*. For example, it might be relatively easy to justify the use of official information from an important agency such as NASA or the Food and Agriculture Organization of the UN but less easy to justify the use of web information produced by another student.

5 Verbatim material should be presented in *quotation marks*, thereby continuing the tradition relating to paper text.

6 Because the copy/paste action is so easy, a student's work could be unbalanced by excessive, downloaded material. In order to make

students more critical, and to control wastage of time, space and paper, *no verbatim material should be more than 10 lines in length*, unless there is very strong justification. In real science, excessive quotation signifies a degree of intellectual laziness.

7 *All images should be fully attributed* to the author's URL and name. For anything which might go on public display (for example in a poster presentation at a conference), permissions for use should have been obtained from the author. Image links should be used sparingly, since downloading could occupy significant amounts of time for a curious tutor.

8 Web sites should have been *visited several times* by the student to show that they are durable. In the list of references to web addresses, the student should give some measure of the stability of the quoted web sites. For example, how often have they successfully visited the site, and over what period?

9 Where possible, when web pages are cited in an assignment, students should *keep electronic copies of these web pages*.

Finally, be prepared to provide proof of visits to web sites. If there are problems, these can prove the validity of what you have done. This evidence comprises *dated* paper copies of web pages.

3.4 SOME ADVICE TO TUTORS

3.4.1 Making Contact with Authors

A web author might not welcome the arrival of hundreds of emails from students, each asking permission to copy information. Teachers should warn students of the need to obtain permission but could then offer to send a single request if many students all want to copy from the same web pages.

Authors of web pages that might prove useful for teaching should be encouraged to add a 'permission to copy' statement with defined restrictions. A possible form of words could be:

All the information listed is believed to be in the public domain. These lists may be freely copied for personal and educational purposes. If you distribute sections for non-profit purposes, please acknowledge the source. We reserve the copyright for any commercial use.

3.4.2 Following up Links in Students' Work

Tutors often set assignments with a word limit – for example, 2000 words or 6 pages. However, if the assignment includes electronic supporting material, what are the responsibilities of the tutor for reading everything? In other words, how do we define the volume of a piece of student's work when it is produced electronically? It would seem reasonable to agree with the students that *tutors will read only the students' own words* and will, within reason but not exhaustively, investigate links. The latter can be a useful source of material for the tutor. In effect, the students are acting as a meta-search engine (see Chapter 5).

3.4.3 Directing Students to Appropriate Web Sites

Increasingly, tutors will have to make teaching materials available to students via the internet. Conversely, tutors will not be able to ignore the internet as a resource and should be prepared to direct students to useful and reliable web sites and pages. It will be incumbent on tutors to present models of good practice in:

- identifying appropriate and reliable web sites,

- citing these web sites in an appropriate way,

- periodically ensuring that the target web sites continue to exist.

The most recent versions of word processing packages such as Word will automatically create hyperlinks from any suite of characters which begins with http://

3.4.4 A Caveat About Exclusion

The major developments of the internet and web were predicated on the idea of *open access*. It is sad that many universities restrict access to the academic content of their web sites, excluding students from outside the institution. It is arguably true that allowing open access benefits an institution by increasing the numbers of visitors to its site and by demonstrating the openness of its pedagogy. Should another institution use the materials, then one might consider that imitation is the highest form of compliment.

3.5   PLACING YOUR REFERENCES IN AN ASSIGNMENT

In the text give the *URL and date*. The full references, listed in alphabetical order, should be given at the end. As a reader and marker, I prefer to see the web references listed separately from other information sources. This is because:

- almost all paper-based resources have been refereed in some way or another. The same cannot be said of web-based resources;

- a reader with access to a library can almost always, eventually, get hold of a paper-based reference quoted in a reference list. This is not true of electronic references. A significant proportion of web-based sites have disappeared within a few weeks of making their appearance;

- because of the previous point, many web references have less academic cachet than paper-based references and should be treated separately;

- the reader may want to follow up web links immediately. This is easier if they are all grouped together.

3.6   SPECIFICATIONS FOR REFERENCES

The following specifications were collected from a university-based web article by M. Quinion at:

http://www.indiana.edu/~diatom/webcite.txt 1/6/00–27/7/01 *which quotes a range of sources.*

3.6.1   Citation of www, FTP, Gopher or Telnet sources

In general, use:

<author's name><title of document><URL><the first and last date the document was inspected by a web site visit>

For example (Fig. 3.4):

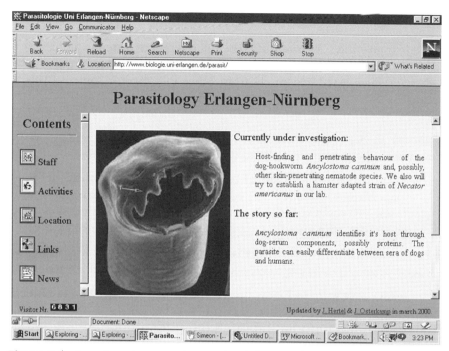

Figure 3.4 Scanning electron micrograph of the mouth of a human parasite, the hookworm

Haas, W. & Osterkamp, J. Hookworm research at Erlangen-Nurnberg University, Germany: http://www.biologie.uni-erlangen.de/parasit/ contents/research/nematod.html 17/11/00–4/4/01

Try to ensure that the URL stays on one line if you can. If it runs onto the next line, *do not use a hyphen*.

3.6.2 Email citation

Email messages should be cited as personal communications (*pers. comm.*). These can be presented in the same format as www sources. For example:

- in the text, the reference would look as follows:
  <pers.comm><name of correspondent> <date>
  For example: (pers. comm., Dussart, 2000)

- in the reference list, the email would look as follows:
  <Name of correspondent> <email address of the correspondent>< Title for the issue discussed><Nature of the communication><email address of the person currently writing><date>
  For example:
  Nicholls, M. (mkf9@cant.ac.uk) Predatory birds. Request for information from Dussart (gbd1@cant.ac.uk) 1.11.2000

This subject of citing on-line material can be followed up by visiting the following web site at Indiana State University. There are many links from this site:

http://www.indiana.edu/~librcsd/eval/citing.html 7/3/01–17/7/01.

3.7 COPYRIGHT AND PLAGIARISM

3.7.1 Introduction

In the following, for ease of use, URL references have been put together in section 3.7.4. In the introduction to this book, I tried to appeal to your motives as a dedicated learner, who is trying to develop your own mind. If you accept the role, you would find plagiarism an anathema.

Copyright infringement and plagiarism are two sides of the same coin. *Plagiarism ('copying') is the theft of language, wording or ideas*, though it can also be considered to be fraud. However, where students commit plagiarism, it is often accidental because they don't know the rules. However, it would be worth your while to know the rules. Plagiarism is a serious misdemeanour in academic institutions and being on the receiving end of a plagiarism panel is a thoroughly unpleasant way to spend an afternoon, even if you are found not guilty.

It is possible to plagiarise images as well as text. Editing by copy/paste makes plagiarism of words and images more likely, and less detectable. At an early stage, you need to be aware of the ethics of plagiarism and Krulwich (1997, URL-1) offers role-play scenarios which examine the ethics of plagiarism and cheating in the context of medical training. Gordon, Simmons and Wynn (1997, URL-2) deal specifically with the definition and avoidance of plagiarism. Using examples, they informatively classify plagiarism into complete-, near-, patchwork-, lazy- and self-plagiarism and offer advice on its avoidance. An excellent review of the *intellectual property rights* situation in the USA is provided by Scheftic (1997, URL-3); much of what Scheftic writes is relevant to Europe.

3.7.2   Copyright

One of the main copyright instruments is the *1886 Berne Convention for the Protection of Artistic and Literary Works*, which was augmented in 1996 by the World Intellectual Property Organization (WIPO) Copyright Treaty and the WIPO Performances and Phonograms Treaty. The 121 signatory states, including the USA, Japan and all the states of the EU, have accepted the principle of copyright for any kind of work, which could include writing, music, software or image production on any kind of medium such as paper, CD, vinyl, magnetic tape, etc. Definitions relating to copyright issues are available at URL-4; for example, there is a comprehensive description of what is meant by a 'Berne convention work'. The WIPO (URL-5) offers support on copyright issues at an international level and there are also national-level agencies such as the Copyright Clearance Center in the USA (URL-6) and the Copyright Licensing Agency in the UK (URL-7). Lean (1995, URL-8) supplies a state-of-the-art review of copyright in a historical context and notes the relatively new acknowledgement of moral rights which, unlike copyright, remain with the author and their heirs for fifty years.

Particular care is needed to give proper attribution for the use of images. In a discussion of the myths of copyright, Templeton (1997, URL-9) notes that copyright is becoming a criminal as opposed to civil issue. Kleiner (1997) observes that scientists are increasingly posting their papers on the internet, even though the copyright belongs to the publishing journal, and suggests that, in future, this might be interpreted as a criminal act.

3.7.3   The Need for Prudence

You should appreciate that

- there might be *a range of views* about a subject, and even more importantly

- the author might have *a hidden agenda*.

In most cases, your tutor does not care whether you agree or disagree with a particular viewpoint but wants you to be intellectually critical. This means that with internet material, you need to use an extremely fine filter with respect to the quality of the material. It is unlikely that an editor or reviewer has checked, revised and filtered the material on your behalf.

So you *must* be prudent. Even primary school children are now using the internet, so this education for prudence needs to begin at a young age. The Copyright Licensing Agency (URL-7) encourages all users of electronic media, including school children, to try to get permission to copy electronic and other materials. Secondly, the development of an attitude of prudence in students needs to be guided and monitored. This means teachers may follow up URL references by the student to web sites; however, many web sites have a short lifespan. Fishman (1995) offers guidelines on software copyright protection and Lee, Groves and Stephens (1997) suggest ways of using electronic media to promote discussion between students and also for monitoring student performance. One approach to this education in prudence is for tutors to set assignments in which students critically compare information obtained from the internet with information obtained from other sources such as books and periodicals.

3.7.4 URL References on Plagiarism and Copyright

URL-1 Krulwich T. A. (1997)
http://www.onlineethics.org/reseth/msindex.html?text
Problems and standards in research ethics. Here are scenarios and role plays relating to situations common in medical schools, presented by the Dean of the Graduate School at Mount Sinai Hospital as part of the WWW Ethics Center for Engineering and Science. Visited 17/1/98–27/7/01

URL-2 Gordon, C., Simmons, P. and Wynn, G. (1997)
http://www.zoology.ubc.ca/bpg/plagiarism.htm
Plagiarism – what it is and how to avoid it. This is a third-year Biology Program Guide, prepared by senior staff in the Faculty of Arts, University of British Columbia. You should pay particular attention to the examples. Visited 17/1/98–27/7/01

URL-3 Scheftic, C. (1997)
http://www.geom.umn.edu/~scheftic/Talks/IPRW/credits.html

Intellectual property rights and wrongs. This is an excellent personal archive which provides URLs for a wide range of source materials. Visited 17/1/98–27/7/01

URL-4   http://www.law.cornell.edu/uscode/17/101.html
United States Copyright Act. Definitions. This comprises information provided by the Legal Information Institute at Cornell Law School. Visited 17/1/98–27/7/01

URL-5   http://www.wipo.org/eng/index.htm
World Intellectual Property Rights Organization. This is the homepage of the WIPO webserver, located at Geneva in Switzerland. Visited 17/1/98–27/7/01

URL-6   http://www.copyright.com
Copyright Clearance Center Online is the homepage of a not-for-profit organisation created at the suggestion of Congress to help compliance with US copyright law. Visited 17/1/98–27/7/01

URL-7   http://www.cla.co.uk
The Copyright Licensing Agency of the UK. This is an official homepage with questions and answers. It stresses that the law is still being made. Visited 17/1/98–27/7/01

URL-8   Lean, M. (1995)
http://ausweb.scu.edu.au/aw95/future/lean/index.html
Copyright and the World Wide Web. Here is a review from the First Australian World Wide Web Conference. Visited 17/1/98–27/7/01

URL-9   Templeton, B. (1997)
http://www.clari.net/brad/copymyths.html
Ten big myths about copyright explained. This is a useful essay expressing personal views and produced by the publisher of an electronic newspaper on the net, with links to other appropriate documentation. Repeated and heartfelt pleas are made by the author not to be contacted for legal advice. Visited 17/1/98–27/7/01

4
Evaluation of Software and Web Sites

When evaluating a web site, there are two concerns, First, *how reliable* is the content and secondly *how easy is it to use*. This chapter gives some guidelines on how to recognise good web sites and then directs you to the types of web site which are academically reliable.

4.1 EVALUATING THE MECHANICAL ASPECTS OF A WEB SITE

Mechanical aspects means the arrangement and presentation of the web site contents. The quality of a web site can be judged on the mechanical aspects and/or the academic content. Two extremes would be:

1 a web site with poor content may be well presented and easy to use. Such a web site might be visited often by the intellectually unwary;

2 a web site with excellent content may be poorly presented and difficult to use. Such a web site could be quite rarely visited.

Here is an *evaluation checklist* to help you to assess the mechanical aspects of a web site or program.

- Is it easy to navigate round the site?

- Are there instructions which are easy to follow?

- Does the software load or download quickly?

- Is the structure conceptually clear?

- Does the software allow you to control precisely what you want to do?

- Does it crash or lock?

- Can you get out of it easily and, on going out, does it take you somewhere sensible?

- Is there any feedback on how you are doing?

- Is there an email address to a webmaster?

- Can you get from the web page on the screen to prior ones on this web site? In other words is there a Homepage button?

Of course, these evaluation points should be taken into account when you are designing your own software or web site.

4.2   DESIGNING YOUR OWN WEB SITE

One of the best ways to recognise a good web site is to be aware of how web sites are designed. Imagine you are designing your own web site material.

It is easy to design your own web pages using relatively few commands from the html language (see Introduction). It is even easier to do so by using special programs such as Microsoft Frontpage or Macromedia's Dreamweaver. Unfortunately, although the latter two programs appear to work in a WYSIWYG way ('what you see is what you get') they add spurious code to the main html document. It can then be difficult to make changes to the raw code later. Dreamweaver is much better than Frontpage in this respect. Undoubtedly, things will get easier in the future as new programs are developed.

No matter which software you use, there are *elementary rules of web page design*. If you are aware of these rules, they will not only allow you to produce your own good quality web pages, but will allow you to evaluate web sites and pages that you encounter on the internet.

1   Understand what you want your web site to do.

2   *Keep it simple* (*KIS*): don't overload the site with flashy objects. (Incidentally, this is also generally true of assignments and examination answers. Speak your truth clearly and simply. *Tutors like to see simple things done well, rather than complicated things done badly.*)

3   Try to ensure that each page is within 3–6 clicks of every other page.

Make it easy for the visitor ('Make it easy for the reader' is a good axiom for assignments and examination answers too).

4 Provide some kind of *index or site map* so that the user knows where they are.

5 Ensure that downloading is quick. Visitors are likely to move to another site if downloading takes more than 20 seconds.

6 Ensure that you *provide an email address* so that you can be contacted about issues on your web site. Be suspicious of any web site that does not do this.

More sophisticated advice includes:

7 Don't use an image format to present text. For example, if you want to put some text on a web page, do it as .html text rather than as text on a .jpg image.

8 Don't use multiple columns of text. Some browsers will represent these as gibberish.

9 Provide alternatives to non-character-based files such as .pdf files; at the very least, provide links to converters which will make them readable to everyone.

4.3 EVALUATING THE INTELLECTUAL CONTENT OF A WEB SITE

As with paper-based materials, a useful criterion of intellectual content is the *nature of the source* of the information. Domain names are a good initial indicator. A new generation of domain names is currently being developed.

The domain name .gov usually signifies reliable government information; be wary, though – governments have been known to lie, especially at times of crisis.

The domain extensions .ac (academic) and .edu (education) are sometimes good pointers. However, it is quite feasible for a student to place their work, such as an essay, on their own university web site. This

Figure 4.1 The web site for the US Agency for Toxic Substances and Disease Registry is a good example of a site which offers clear contact opportunities if they should be needed

essay could then be cited by students in other institutions who are not aware that this is the work of just another student. If you are suspicious, it is often possible to contact the author by email and confirm their status. Beware of any web site which does not allow you to contact someone in authority (Fig. 4.1).

4.4   WORKING BACKWARDS THROUGH A WEB ADDRESS

Two of the major problems you will encounter with web sites are that:

- the web site is still available, but not the particular pages in which you are interested. You may have looked at them some time ago, but now they seem to have disappeared from that web site;

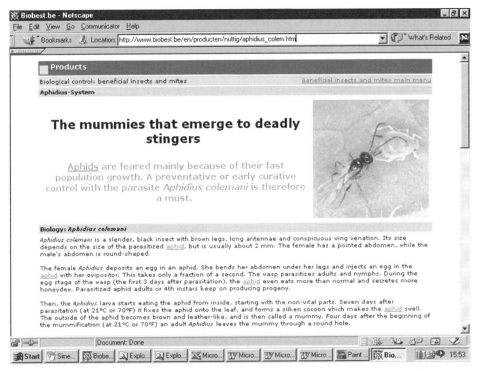

Figure 4.2 *Step 1* http://www.biobest.be/en/producten/nuttig/aphidius_colem.htm

- you have found an interesting web page but you cannot identify its provenance. In other words you don't know what or who, exactly, it belongs to. Can you trust it?

These questions can be addressed by *dissection* of the URL. The principle is to cut off the various parts of the address from the right-hand side until more information about the host is revealed. Thus in section 2.8.4, Fig. 2.11, by removing the various parts from the right-hand side, you would move up the folder structure. Try using your browser on that example to confirm that you understand the process. The procedure is as follows:

http://www.cant.ac.uk/depts/acad/science/sci.htm
http://www.cant.ac.uk/depts/acad/science
http://www.cant.ac.uk/depts/acad
http://www.cant.ac.uk/depts
http://www.cant.ac.uk

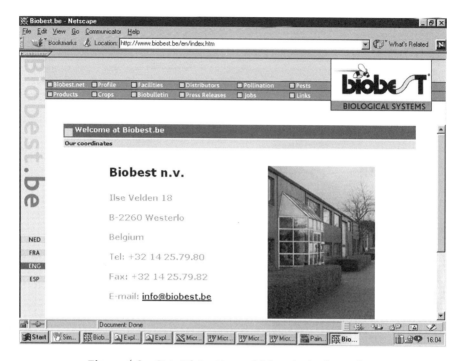

Figure 4.3   *Step 2* http://www.biobest.be/en/home.htm

Here is a more biologically oriented example. Assume it is necessary to find information on *insects as pests*; a colleague has supplied a web address which might act as a starting point:

http://www.biobest.be/en/producten/nuttig/aphidius_colem.htm
15/1/01–17/7/01

In Step 1 (Fig. 4.2), the relevant page is brought up. The starting address yields a web page concerning an insect called *Aphidius colemanni.* The page has lots of valuable information concerning the biology of this parasitoid. However, the page gives no clues as to the provenance or aims of the web site, and it doesn't have a 'return to homepage' link, though the copyright link at the bottom of the page does lead to a postal address.

In Step 2 (Fig. 4.3), the last part of the web address is deleted to see what can be discovered about this site. Thus, the part of the following address which follows the country code Be (meaning Belgium) is removed. (The

Figure 4.4 *Step 3* http://www.biobest.be/en/plagen/default.htm

'eng' part is ignored since it almost certainly refers to a translation into English.)

http://www.biobest.be/en/producten/nuttig/aphidius_colem.htm

This leads to the introductory page of a Belgian firm called 'Biobest Biological Systems'.

Step 3 (Fig. 4.4) involves clicking the button pointing to *pests*, where there are clear images and good biological information.

Step 4 (Fig. 4.5) involves going back one step. If I go back to Step 3 and click on *products*, an index is presented.

Step 5 (Fig. 4.6) comprises scanning down this list for something useful; clicking on *beneficial insects and mites* leads to a list of products, in other words, the predatory insects which are marketed by this company. Clicking on *Aphidius colemani* leads back to the original starting point, i.e. the screen shown in Step 1.

The provenance and aims of this web site are now clear and accessible.

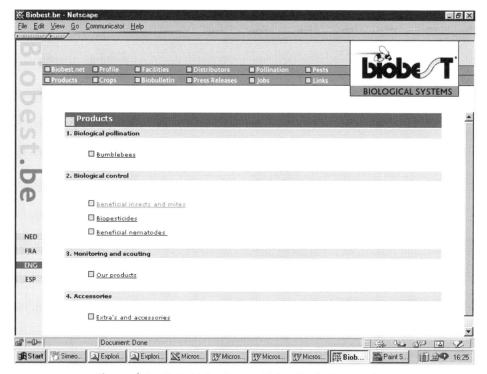

Figure 4.5   *Step 4* http://www.biobest.be/index_eng.htm

This procedure is also useful if you try to visit the site and are told it is no longer available (Fig. 4.7). You are now into . . . *last chance procedures*:

- Often, the site has been moved to another place on the host server. Keep *serially cutting* off the URL until you arrive at the host server. During this process, you might be lucky enough to find the lost pages.

- Many institutional web sites now have a *search engine* for the site. You can use this to search the institution for the information which used to be there.

If you are still unsuccessful, there are three final options, in decreasing order of usefulness:

1 During the URL dissection, you may have come across the name of someone who is responsible for material presented on that site – for example, a tutor. You could *email them directly* to find out whether you can still get the information.

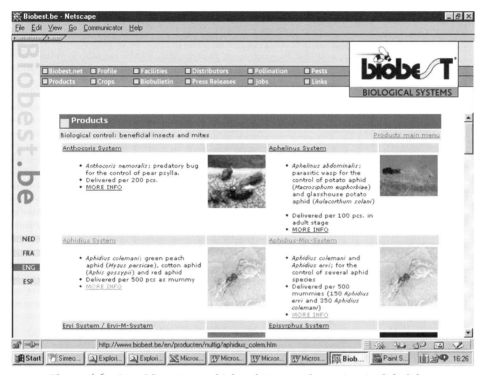

Figure 4.6 *Step* 5 http://www.biobest.be/en/producten/nuttig/default.htm

2 On the homepage of the web site, there may be an invitation to *Contact us*. (All web sites should have one of these but not all do.) Selecting the 'Contact us' link will usually take you to an email address for your query.

3 There may be a hyperlink to the person who designed the site (the *webmaster*). While they will almost certainly have nothing to do with the content, they may be able to point you in the right direction. Email them.

Still unsuccessful? There are other pebbles of information on the intellectual beach. Look for something new and *don't waste any more time on this fruitless search*.

4.5 HIDDEN URL WEB ADDRESSES AND HOW TO GET ROUND THEM

Sometimes, because of the way a web site has been set up, there may appear to be a single URL for which there are several different web pages.

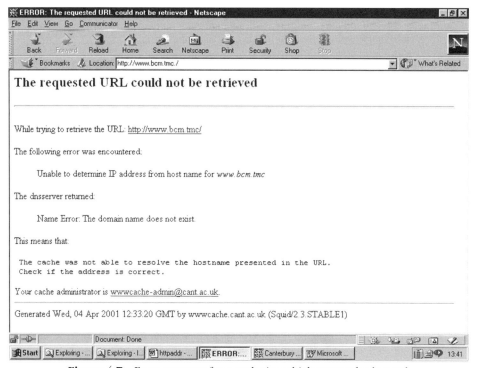

Figure 4.7   Error message for a web site which cannot be located

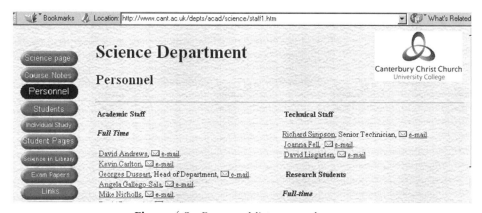

Figure 4.8   Personnel list on a web page

This is usually due to the way frames have been used on the web site. As you move from web page to web page, the URL in the location bar does not appear to change. For example, Fig. 4.8 shows a personnel list.

When the hyperlink for any person is clicked, you are taken to their

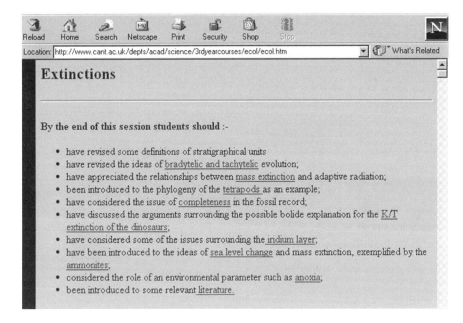

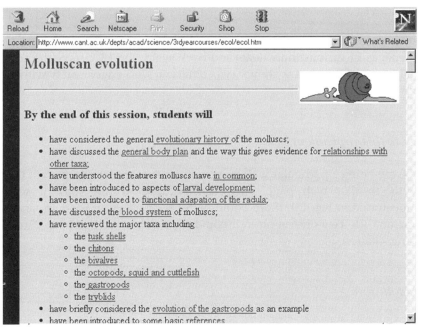

Figure 4.9 Two different web pages, each with the same URL

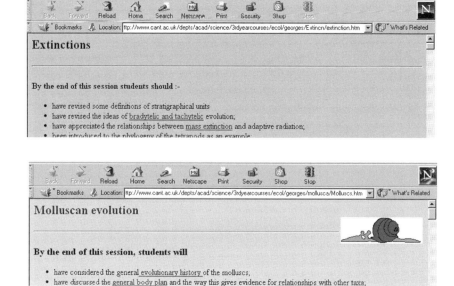

Figure 4.10 The same two web pages, now with their full URLs

homepage. However, as you pass to and fro, from personnel page to the homepages of various individuals and back again, the URL stays the same. To illustrate this, Fig. 4.9 shows two different web pages, each with the same URL.

Why is this a problem?

If you are citing a particular web page, you or your reader might want to go directly to that web page. However, because of the kind of set-up demonstrated in Fig. 4.9, there could be 50 web pages all with apparently the same URL. To a critical tutor, it could appear that you had been careless in your citation.

The solution

With Netscape, the answer is to click on the major part of the page using the right mouse button. A drop-down menu appears; select *open frame in new window*. This will reload the page with its complete URL address in the location bar, as shown in Fig. 4.10. Unfortunately, Internet Explorer does not seem to offer an equivalent option.

4.6   ACADEMICALLY RELIABLE WEB SITES

Now you know the mechanics of navigating round web sites, we need to consider their *intellectual and academic worth*. How do you know that the pages you are looking at are worthy of citation?

4.6.1   Academic Journal Web Sites

Journal web sites are reliable but are usually only fully open to subscribers. However, it is often possible to see abstracts and even to do searches.

A search web site based on the keyword *mitochondria* on the *Journal of Experimental Botany* web site at:

http://jexbot.oupjournals.org/search.shtml 3/7/00–19/7/01

yielded 78 articles, which could be viewed as abstracts, or downloaded as 'full text and image' .pdf files (Fig. 4.11). This web site is an excellent source of botanical information.

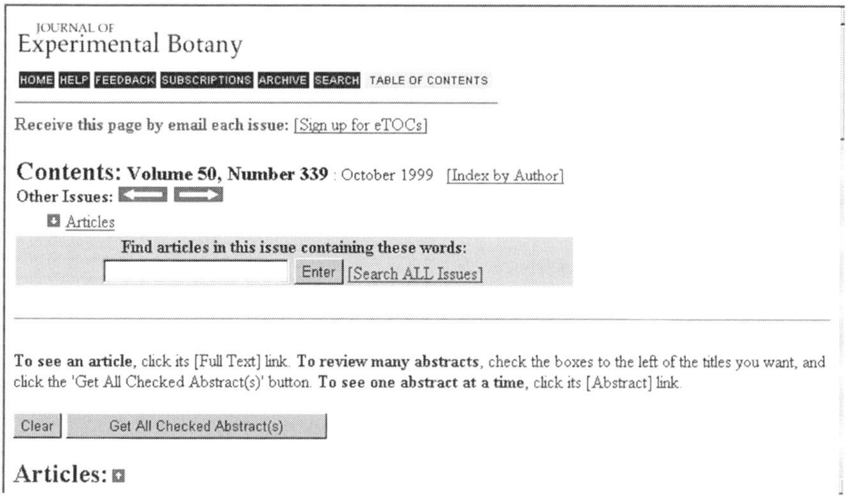

Figure 4.11   Web search-page of the *Journal of Experimental Botany*

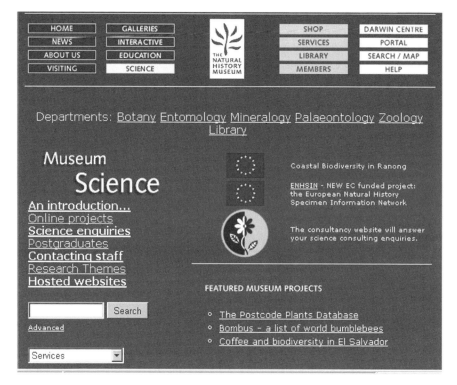

Figure 4.12   Welcome page of the Natural History Museum in London

One of the best routes to bioscience academic journals is via the *BioMedNet* web site at:

http://journals.bmn.com/journals 3/7/00–19/7/01

4.6.2   Museum Web Sites

Web sites which are run by *museums,* such as the Natural History Museum (Fig. 4.12), are reliable:

http://www.nhm.ac.uk 3/7/00–18/7/01.

Their immediate sources of information are often directed at children and the lay public. They can therefore appear to be limited, though searching more deeply can yield good information. A search with the keyword *sharks* yielded 47 items including:

- *'Could we still study isolated primitive shark teeth without a SEM?'* which was an abstract from a Palaeontological Association Conference at Birmingham (1996),

- a link to the Smithsonian Institution in Washington DC where there was a freely available key to fossil shark teeth.

4.6.3   Learned Society Web Sites

Learned societies such as the Institute of Biology (http://www.iob.org 3/7/00–17/7/01) provide reliable information. A list of fact sheets is available through the web site on a monthly basis and a full text version of the current issue of the Institute journal *Biologist* is also available on-line.

The Royal Society (http://www.royalsoc.ac.uk 3/7/00–17/7/01) is the premier scientific institution in the UK (Fig. 4.13) and provides access to a useful range of abstracts from its journals. There is an efficient search engine for the site.

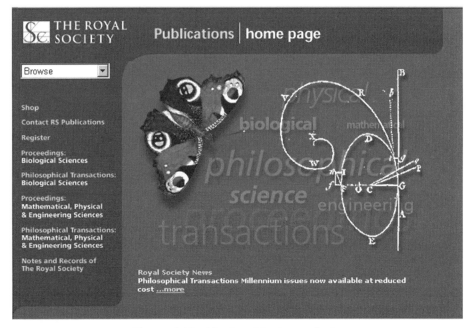

Figure 4.13   The Royal Society web site

4.6.4   *Newspapers, Magazines and News Services*

On-line *newspapers and news services* often provide useful information. These services tend to represent secondary sources. In other words, they usually comprise articles written by journalists rather than articles written by the researchers themselves. The pattern of presentation of such articles is usually to collate the work of an individual or team, and then to consult other experts in the field who are asked for their opinion of the research in question. This constitutes a form of quasi-peer review. However, it is not remotely as rigorous as the peer-review systems for journals such as the *Journal of Experimental Botany* or the *Proceedings of the Royal Society.*

Broadsheet, quality newspapers such as *The Guardian* (http://guardianunlimited.co.uk 3/7/00–17/7/01) are useful targets for serendipitous scientific information searches.

New Scientist is a good example of a science magazine (http://www.newscientist.com 3/7/00–17/7/01). The magazine has high production values and is renowned for the quality of the artwork which it has presented alongside its science over the years. Interesting information

Figure 4.14   Web page of the *New Scientist* magazine

can be found serendipitously by looking around the *New Scientist* web site (Fig. 4.14) but, unfortunately, one must be a subscriber to get access to the detailed articles. Once access has been permitted, however, previous issues covering a period of ten years can be searched. It is certainly worth taking out a student subscription to have access to this facility.

New Scientist also offers a free web-link service; web pages are reviewed in a useful and informal way. For example, at:

http://www.newscientist.com/weblinks

there are links to over 1600 science sites. Each link is accompanied by a relatively detailed commentary.

The Scientist is a news magazine which makes good materials accessible over the internet (Fig. 4.15). It offers a newsgroup service which sends, to your inbox, monthly emails containing .html addresses of recent articles. You can subscribe to the service for free at:

http://www.the-scientist.com/homepage.htm 3/7/00–17/7/01

4.6.5   Broadcasting Institutions

Broadcasting institutions such as the BBC are another valuable source of reliable information. The BBC has an excellent, serendipitous site with an efficient search engine:

http://newssearch.bbc.co.uk/default.stm 3/7/00–17/7/01

A search under the heading of Science/Technology using the keyword '*E. coli*' (a common gut bacterium in mammals) yielded 142 hits. Many of these dealt with a serious outbreak of food poisoning which had been completely unreported in the national newspapers.

4.6.6   Review Journals

The Elsevier *Trends* series of review journals is an excellent source of reliable, refereed information. The series includes *Trends in*

Biochemical Sciences

Biotechnology

Figure 4.15   Web page of 'The Scientist' on-line news magazine

Cell Biology

Cognitive Sciences

Ecology & Evolution

Endocrinology and Metabolism

Genetics

Immunology

Microbiology

Molecular Medicine

Neurosciences

Parasitology

Pharmacological Sciences

Plant Science

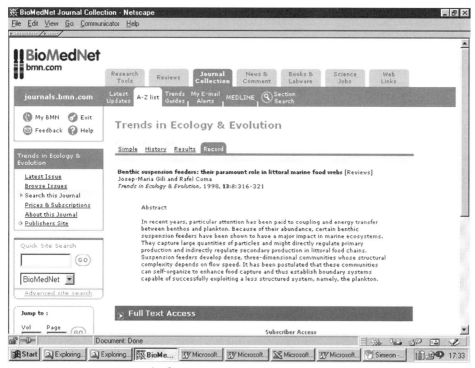

Figure 4.16   Typical result of a BioMedNet search

Access to these journals is available through the *BioMedNet web site*. Abstracts are available for all the articles, and full texts can be purchased on-line (Fig. 4.16) at:

> http://journals.bmn.com/journals/list 5/3/01–27/7/01.

4.6.7   *Bibliographic On-line Services*

These are the major resources used by academic scientists at all levels.

PubMed

Established in 1988 as a national resource for molecular biology information, the US National Center for Biotechnology Information (NCBI) disseminates *biomedical* information. You can get access (Fig. 4.17) at:

> http://www.ncbi.nlm.nih.gov/About/contactinfo.html

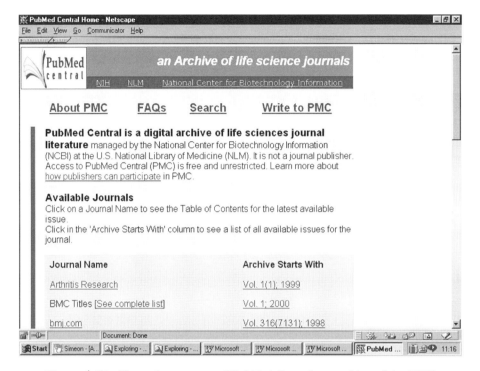

Figure 4.17   The welcome page of PubMed Central, an archive of the NCBI

In particular, it operates two valuable open-access databases:

- *PubMed* offers access to over 10 000 of MEDLINE's biomedical journal citations.

- *PubMed Central* is an archive of biosciences journal articles, with unrestricted access to 49 life sciences journals.

To give you a 'feel' for this database you might consider the following. A search in PubMed Central using the keyword *operon* yielded 110 articles. Using the keywords *lac AND operon*, there were 31 articles covering a period from 1996 to 2001. A search using the keywords *foliar AND cineole content* gave no positive hits; nor did *seed AND dispersal AND halophytes*. This implies that, as one might expect, the archive is particularly oriented towards the molecular and medical subjects. However, reducing the keyword list to *seed AND dispersal* yielded 3 wide-ranging hits from the *Proceedings of the National Academy of Sciences*:

- Chloroplast DNA footprints of postglacial recolonization by oaks

- Directed seed dispersal by bellbirds in a tropical cloud forest

- *Centaurea corymbosa*, a cliff-dwelling species tottering on the brink of extinction: A demographic and genetic study

One of the most significant advantages of these archives is that full-text versions can be downloaded at no cost. These articles are hyperlinked; so, for example, clicking on the thumbnail diagrams in the left margin reveals the full-sized image (Fig. 4.18). The article concerned is Tamames, J. (2001) Evolution of gene order conservation in prokaryotes *Genome Biology* 2(6).

Web of Science

Web of Science is organised by the Institute for Scientific Information (Fig. 4.19) and access is available at:

http://wos.mimas.ac.uk 1/3/99–17/7/01

The system uses sophisticated protocols to search for articles according to the keywords you supply. The most important articles are noted first, and become decreasingly irrelevant as you proceed through the list of results. It is possible to *mark* the most relevant articles, obtain a summary list, and send this list to your email inbox, or print it out, or send it to your own bibliographic software (e.g. Procite).

For example, suppose you are interested in the drag coefficients of shells of invertebrate animals. You therefore use the keywords *drag AND shells*. The program highlights four articles in the period 1991–8 (Fig. 4.20). This seems to be rather few, so you therefore select the most relevant article and double-click on it.

This gives the details of the paper and an abstract. You now click on the button which offers related records, and are given 100 references. To illustrate the order of relevance, the title of the first article on each page is now listed.

1 The hydrodynamic characteristics of 6 scallop species . . . most relevant

2 Size-related hydrodynamic characteristics of the giant scallop

3 Tuned oscillations in the swimming scallop

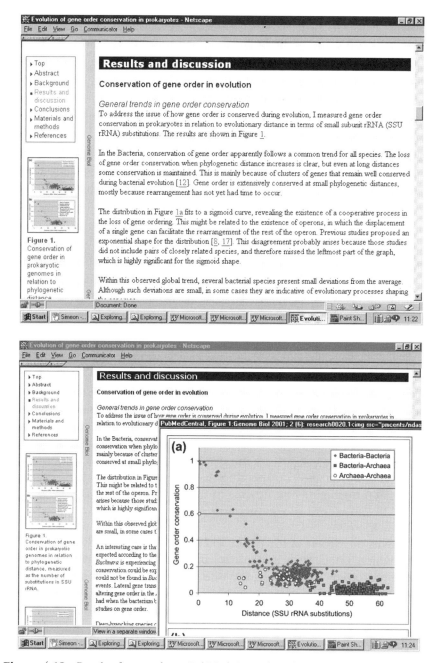

Figure 4.18   Result of a search on PubMed Central, and magnification of one of the figures in the article

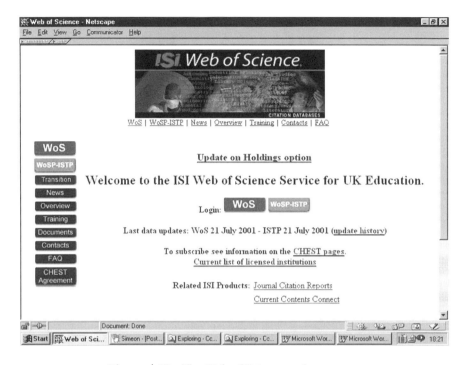

Figure 4.19 The Web of Science welcome page

4 Dynamics and energetics of scallop locomotion

5 Some aspects of the functional-morphology of the shell of infaunal bivalves (mollusca)

6 Hydrostatic stability of fish with swim bladders – not all fish are unstable

7 The biology of oniscid woodlice of the genus *Tylos*

8 Design of the predatory legs of water bugs

9 Metabolic responses to sulfur in lucinid bivalves

10 Noise generated by the jaw actions of feeding fin-whales

11 Energetic efficiency and ecology as selective factors in the saltatory adaptation of prosimian primates

12 *Maida nov gen*, the oldest known nektoplanktic bivalve from the Pridoli (Silurian) of Europe

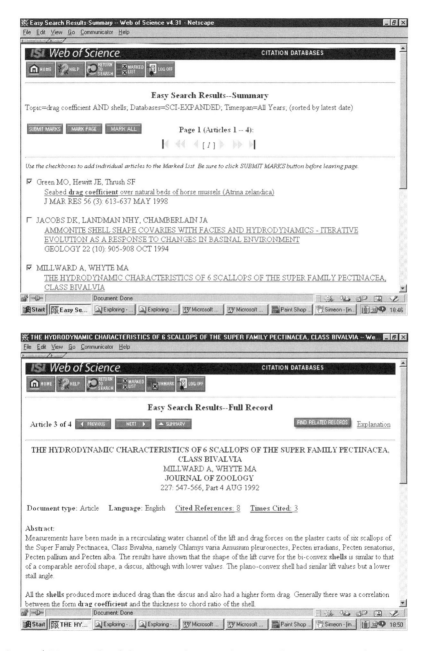

Figure 4.20   Details of the most relevant reference. The top picture shows the first result of the search. The button for related articles is at the top right of the bottom picture

13 Phenetic discrimination of biometric simpletons: paleobiological implications of morphospecies in brachiopods

14 *Oelandiella*, the earliest Cambrian helcionelloid mollusc from Siberia . . . least relevant

It is now possible to order the *full text* of the most relevant references through the university library and the inter-library loan system. You may have to be parsimonious, since it might cost money to obtain the article.

However, if you are working in a British university, your university library will almost certainly have negotiated access possibilities such that you can download some full-text articles, for example through ISTP (the Index of Scientific and Technical Proceedings), IDEAL (the International Digital Electronic Access Library), Science Direct Web Editions, Blackwells Science Internet and Web of Science (formerly BIDS). Before opening your wallet, *make thorough enquiries* at your university library enquiry desk.

Alternatively, you can investigate whether the article is available on-line for free through BioMedNet.

4.6.8 Companion Web Sites

These are web sites which are meant to be used *in association with a textbook or CD*. In some countries, there is a tradition of having a teaching curriculum which is based on the content of a specific text. This is called 'teaching to the text'. It tends to be frowned on as a pedagogic method in the UK, where students are expected to consult as wide a range of the literature as possible. However, as the quality of textbooks increases, and their relationship with the academic literature becomes blurred, and as web links develop, teaching to the text may become the norm. Here are some examples of *companion web sites*.

Example 1

This is a book which might be used by a student at school or in a foundation or first year at university. It is associated with the textbook:

Purves, W., Sadava, D., Orians, G. and Craig Heller, H. (2000) *Life: The Science of Biology*, 6th edition, Sinauer Associates, Inc. and W. H. Freeman and Company, Sunderland, Mass. 1044 pp.

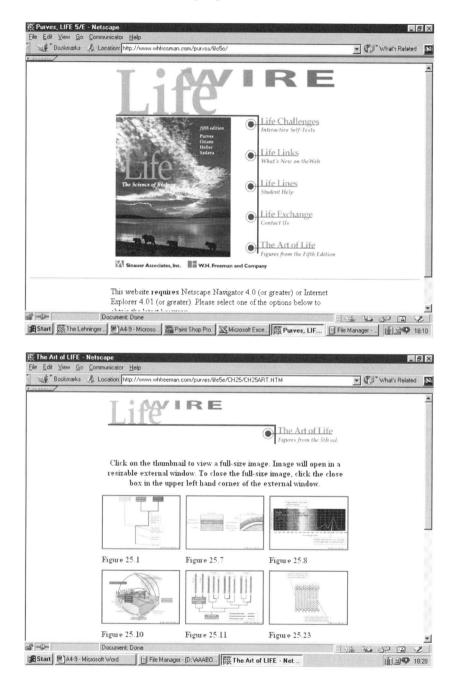

Figure 4.21   Companion site for a first-year undergraduate biosciences textbook

The web site at:

> http://www.whfreeman.com/purves/life5e/index.html#top
> 6/6/01–27/7/01

offers a range of possibilities, including question-and-answer contact with the authors, on-line tutorials on maths and student survival and downloadable figures from the text (Fig. 4.21).

Example 2

Another commonly used foundation textbook with a companion web site is:

> Raven, P. and Johnson, G. (2000) *Biology*, 6th edn, McGraw-Hill, NY.
> 1310 pp.

The web site (Fig. 4.22) is at:

> http://www.mhhe.com/biosci/genbio/rjbiology 30/6/01–27/7/01

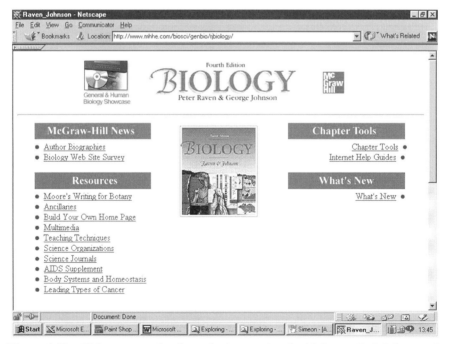

Figure 4.22   Welcome page for the web site associated with Raven, P. & Johnson, G. (2000) *Biology* 6th ed. McGraw-Hill, NY

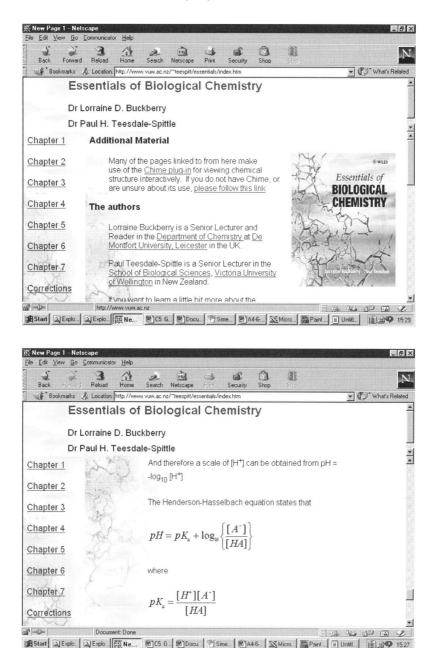

Figure 4.23 The companion web site for Buckberry, L. & Teesdale, P. (2001) *Essentials of Biological Chemistry.* Wiley, Chichester

There are a few problems here; for example, the publisher's web search engine couldn't find the book. This web site is less comprehensive than for Purves *et al.* but does offers a range of possibilities, including answers to questions posed in the paper textbook. One of the most useful attributes is the list of web sites associated with each chapter. For example the chapter on gene technology (http://www.mhhe.com/biosci/genbio/rjbiology/chap19.html) directs the reader to seventeen web sites.

A particularly useful page on this web site offers links to a range of web sites which offer help on using the internet:

http://www.mhhe.com/biosci/genbio/rjbiology/internet.html
30/6/01-27/7/01

Example 3

A good example of a companion web site associated with a more specialised textbook (Fig. 4.23) is:

Buckberry, L. and Teesdale, P. (2001) *Essentials of Biological Chemistry*, Wiley, Chichester, UK. 238 pp.

The website is at:

http://www.vuw.ac.nz/~teespitt/essentials/index.htm 5/7/01-27/7/01

4.6.9 *Freely Available Back Issues of Major Journals*

HireWire Press, hosted by Stanford University in the USA, offers free access to major journals which are usually over 12 months old. Most undergraduates will not need to work at the absolute cutting edge of the research process, so these journals could be a valuable resource for learning some of the background.

Table 4.6.9 lists the biologically relevant journals which are available on this web site. *Proceedings of the National Academy of Sciences* and *Science* would probably be the most useful for novice biologists.

The University of Buffalo library in the USA is another site which offers access to some full text and other journals. It is located at:

http://ublib.buffalo.edu/libraries/units/sel/collections/ejournalfulltext.html 23/7/01

Figure 4.24   The welcome page of *HighWire Press*, Stanford University Libraries in the USA

The site has a long list of journals, some of which link to journal web sites where free downloads of full text are possible. For example, it was possible to download the following article as .pdf files:

Jimenez M. *et al.* (1997) Mycotoxin production by *Fusarium* species isolated from bananas. *Applied and Environmental Microbiology*, **63**, 364–369.

Table 4.6.9 Biologically relevant journals which are freely available on the HighWire web site, albeit slightly out of date

Academic Medicine	*Journal of Experimental Medicine*
Advances in Physiology Education	*Journal of General Physiology*
Age and Ageing	*Journal of General Virology*
American Journal of Botany	*Journal of Histochemistry and*
American Journal of Clinical Nutrition	*Cytochemistry*
American Journal of Physiology	*The Journal of Immunology*
Antimicrobial Agents and	*Journal of Lipid Research*
Chemotherapy	*Journal of Molecular Diagnostics*
Applied and Environmental	*Journal of Neurophysiology*
Microbiology	*Journal of Neuroscience*
Archives of Disease in Childhood	*The Journal of Nutrition*
Arteriosclerosis, Thrombosis & Vascular	*The Journal of Physiology*
Biology	*Journal of the Royal Society of Medicine*
BMJ	*Journal of Virology*
Biology of Reproduction	*Journals of Gerontology Series A:*
Chemical Circulation	*Biological Sciences*
Clinical and Diagnostic Laboratory	*Learning & Memory*
Immunology	*Microbiology*
Clinical Microbiology Reviews	*Microbiology and Molecular*
Development	*Biology Reviews*
Drug Metabolism and Disposition	*Molecular and Cellular Biology*
The EMBO Journal	*Molecular Biology of the Cell*
Endocrine Reviews	*Molecular Biology and Evolution*
Experimental Biology and Medicine	*Molecular Endocrinology*
Genes & Development	*Molecular Interventions*
Genetics	*Molecular Pharmacology*
Genome Research	*The New England Journal Medicine*
Imaging	*News in Physiological Sciences*
Infection and	*Nucleic Acids Research*
International Journal of Epidemiology	*The Oncologist*
International Journal of Systematic and	*Pharmacological Reviews*
Evolutionary Microbiology	*Physiological Genomics*
Journal of Applied Physiology	*The plant cell*
Journal of Bacteriology	*Plant Physiology*
Journal of Biological	*Proceedings of the National Academy of*
Journal of Cell Biology	*Sciences*
Journal of Cell Science	*RadioGraphics*
Journal of Clinical Endocrinology &	*Radiology*
Metabolism	*RSNA Index to Imaging Literature*
Journal of Clinical Microbiology	*Science*
Journal of Experimental Biology	*Stem Cells*
Journal of Experimental Botany	*Tobacco Control*

4.7   *What Kinds of Literature Sources will Earn the Best Credit with Tutors at Undergraduate Level?*

My personal ranking system would look like Figure 4.25, starting off with most creditable. All would, of course, have to be appropriately cited. Other tutors might have their own preferences.

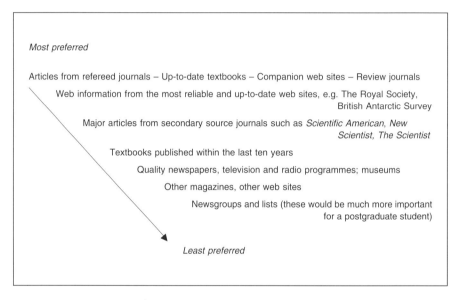

Figure 4.25   Ranking system of literature sources

5
Search Engines

Search engines are *programs* which are used to look for information on the internet. They use different kinds of protocols and criteria for doing their job. Some are much more effective than others for doing academic searches. No two search engines are the same, so consider using two or three engines, even if the first one appears to have been successful. You may find extra information by extending the search.

5.1 TYPES OF SEARCH ENGINE

There are three types of search engine – human-driven, robotic and metacrawlers.

- *Human-driven*
  An example is Yahoo. Human editors classify web sites into categories. This type of engine is best for general enquiries.

- *Robot-driven*
  Here the robot is a program called an 'automated software crawler'. An example is AltaVista. A robot program crawls over the internet picking up the text from web page metafiles. When you interrogate the crawler, it searches for a match in the information it has collected. This type of engine is best for specific enquiries.

- *Metacrawlers*
  These forward your query to several other search engines at once.

5.2 THE LOGIC OF A SEARCH

You can use a hyphen to represent a minus sign. In this way you can subtract or remove things you don't want to include. For example, suppose you want some information on *flowers* for an assignment on botanical aspects of horticulture. If you search for 'flowers' and keep getting the commercial flower delivery service called Interflora you can search for flowers *minus* Interflora: *Flowers-Interflora*.

Boolean searching

There is a special procedure which is used for constructing search phrases based on *Boolean logic*.

The terms AND, OR, NOT are used singly or collectively to connect the words in a search.

- *OR* is a default. If you use two words, the search will yield articles which contain either one of these words. Use of the words '*flavoprotein phytochrome*' will have the same effect as using the words '*flavoprotein OR phytochrome*'. Using this OR term, Google found 14,500 articles, for example, the article at:

  http://www.beri.co.jp/scop/data/scop.1.005.006.000.000.html 22/7/01
  comprised a structural classification of proteins from the MRC Laboratory of Molecular Biology and Centre for Protein Engineering hosted by the Biomolecular Research Institute, Osaka

- *AND* limits the search results to articles that contain all terms that are connected by AND. For example, a search for *flavoprotein and phytochrome* will return all articles that contain both of these words. On this basis, Google found 143 articles, for example the article at:

  http://www.wm.edu/biology/bio419/outline3.htm 22/7/01
  comprised a subject outline for Photophysiology from a plant physiology course at the College of William and Mary in the USA

- *NOT* is used to exclude articles containing certain terms. For example, you would use this if you wanted to search for articles about *flavoprotein* that did not relate to photosynthesis. When the keywords *flavoprotein NOT photosynthesis* were used in the normal search routine, Google found 193 articles, almost all of which featured photosynthesis. However, when the advanced search option was selected using the same keywords, Google came up with 8020 articles! The first was:

http://invader.bgsm.wfu.edu/overview.html 22/7/01
and comprised a description of research into the biochemistry of four unusual flavoproteins in Gram-positive bacteria, studied at Wake Forest University in the USA

By using several of these Boolean words, complex and specific searches can be developed. If you are searching for a *phrase*, the best search engines will look for exactly that phrase if it is defined by *quotation marks*; for example "natural selection".

5.3   SEARCH ENGINES COMPARED

Five search engines were tested by a search related to stomata. In particular, I wanted relatively recent information on *stomatal control* but not detailed, esoteric research papers. This kind of comparison is a useful exercise for any user to undertake. It gives you a feel for the character of the search engines and you will probably develop an affinity for one or two. The following research took place in October 2000. Slightly different results might be expected now.

Yahoo   http://uk.yahoo.com

236 hits

Yahoo is a human-based system which has both UK and worldwide search capabilities. This is a vintage search engine. It is investigated in some detail here to show how the academic reliability of a site can be investigated by truncating the URL address (see section 4.4).

• Yahoo has a biology section. A search through the following folder structure got nowhere since it leads only to lots of pages to do with flowers: Home > Science > Biology > Botany >

• An advanced search using "*stomatal control*" got 236 hits. At least one of these looked promising:

    http://agnews.tamu.edu/drought/drghtpak98/drght46.html   31/10/00–27/7/01

- The page gave lots of apparently good technical and readable information but I couldn't easily see what it belonged to. So, I serially truncated the web address to see where I might end up.

  - http://agnews.tamu.edu/drought/drghtpak98 *Texas drought management strategies*

  - http://agnews.tamu.edu/drought *The Texas Agricultural Extension Service and Texas Agricultural Experiment Station*

  - http://agnews.tamu.edu *Texas A&M University Agriculture Programme*

This was evidently a university site. An investigation of the homepage showed lots of information concerning agriculture which could be useful in a later assignment. I bookmarked it (see section 2.7).

Altavista   www.altavista.co.uk

282 hits

Altavista, now dead, was a web crawler and was useful for unusual or specific searches. A search for *stomata* went through the following levels:

- Home/New/Library/Science/Biosciences/Botany/Plant Biology (14) and then directed me to 14 web sites, some of which were good university sites but none of which seemed as if they would tell me anything about stomata.

- An advanced search yielded 282 hits including a paper on the importance of actin filaments in stomatal control by Korean authors and published in the journal *Plant Physiology* (http://www.postech. ac.kr/~ylee/paper/paper6.html). It was a little too specialised for my objectives. I would have had problems in interpreting the homepages at http://www.postech.ac.kr/~ylee since they are written in Korean.

Google   www.google.com

337 hits

My favourite search engine, Google, is an excellent crawler with a special ranking protocol which seems particularly useful for finding academic information. The development of the protocols started in 1995 and Google now handles 50 million searches per day.

In a search for *stomata* there were 16 000 hits in 0.09 seconds. An advanced search had to be done to whittle these down. Google found 337 in 0.6 second. One of these was an excellent site, though the illustrations need working on. It was appropriate for both teachers and students and was offered by a university lecturer at:

http://www.zoo.utoronto.ca/able/volumes/vol-13/3-brewer/3-brewer.htm
31/10/00–27/7/01

Excite   www.excite.co.uk

40 hits

Excite was a crawler-based system, now defunct, with both UK and worldwide capabilities. It was moderately successful when searching for academic subjects. A search for *stomata* gave 40 hits; some were exactly what was being looked for. For example, photographs of stomata were found at Cornell University in the USA:

http://biog-101-104.bio.cornell.edu/BioG101_104/tutorials/botany/
leaves.html 31/10/00–27/7/01

Excite also directed me to the University of Hamburg which offered an excellent English-language site, including diagrams. It described all the basic biology and biochemistry of stomata, at:

http://www.rrz.uni-hamburg.de/biologie/b_online/e32/32f.htm
31/10/00–27/7/01

Askjeeves   www.askjeeves.com

6 hits

This is a human-organised directory which answers direct questions posed in plain language. A search hierarchy of Ask Jeeves Home > Science >

Biology > Botany > Plant_Physiology led to a number of web articles, none of which looked particularly relevant. A search with the keyword *stomata* yielded only 6 hits, though it did direct enquiries to other relevant sites. It also showed how relevant the cited pages were for similar queries (not very). The specific search for *stomatal aperture* resulted in a recommendation to visit *Encyclopaedia Britannica*. This was rather unsatisfactory since I could have gone there in the first place. There was, however, an option to post the query for general help.

5.4 LIST OF OTHER SEARCH ENGINES

Use the starred ones first.

- *alltheweb****   www.alltheweb.com
  Good sophisticated search engine that brings up scientific pages. This is a good crawler for unusual addresses.

- *Go2Net*   www.go2net.com
  Queries lots of other search engines at the same time.

- *Hotbot***   www.hotbot.lycos.com
  Has a Science and Technology option.

- *LookSmart*   www.looksmart.co.uk
  Human-powered system that classifies sites both in the UK and worldwide. There is no obvious scientific dimension.

- *Lycos*   www.lycos.co.uk
  Crawler-based system offering UK and worldwide possibilities. It subsumed Excite.

- *Northern Light *****   www.northernlight.com
  Crawler-based system which often reveals hard-to-find information and is popular with researchers.

- *SavvySearch*   www.savvysearch.com
  Metacrawler.

- *SearchUK*   www.searchuk.com
  Crawler-based system for the UK only.

- *Snap*   www.nbci.com
  No obvious scientific dimension.

- *Surfwax******   www.surfwax.com
  This is a sophisticated, fast metacrawler. Probably good to use if you are stuck.

- *UKPLus*   www.ukplus.co.uk
  A human-powered service for the UK; not much use for academic things.

- *UKMax*   www.ukmax.com
  Crawler-based system offering UK and worldwide possibilities.

- *VirginNet*   www.virgin.net
  This is a special edition of Google which is able to focus on the UK.

5.5 ANALYSING THE RESULTS FROM A SEARCH ENGINE (GOOGLE)

Google is useful because it lets you see the whole URL and by judiciously reading it, you can get a good feeling for the reliability of the web site. For the example shown in Fig. 5.1, I wanted some technical information on molluscan shells, so I used the search word *conchology*. In less than 2 seconds, Google found over 400 web pages. Unfortunately, the search word was too general, and threw up shell clubs, which were not relevant to my interest.

A more precise search using the phrase *engineering molluscan shell architecture* yielded a long technical article at http://www.nas.edu/bpa/reports/bmm/bmm.html 4/4/01–27/7/01, with figures and references, and which fulfilled the objectives perfectly.

> ... studying how nature accomplishes a task, or how it creates a structure with unusual properties, and then applying similar techniques in a completely different context or using completely different materials. An example of this approach is the study of the laminated structure of clam shells, which has been reverse-engineered to design a metal ceramic composite twice as strong as other composites and an order of magnitude tougher, and constructed of more robust materials than its natural analogue.

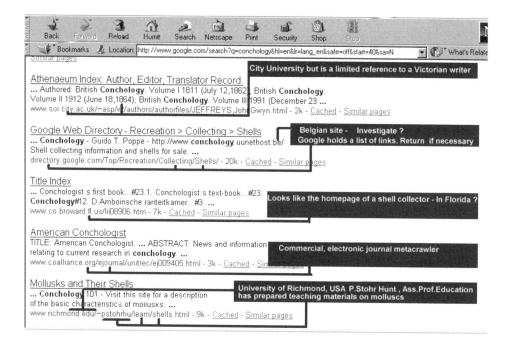

Figure 5.1   Interpretation of a Google search result

Figure 5.1 partly demonstrates that the more you look at URLs, the more of a 'feel' you have for what they mean. The acquisition of this skill can mean that you no longer rely on a combination of keywords and serendipity (lucky finds) but can *hone the search* by your own craftsmanship and familiarity with the system.

5.6   PROBLEMS WHEN THE SEARCH ENGINE DOES NOT REVEAL THE URL OF THE TARGET SITE

This is a problem that arises with the frames that the web designer uses to construct the pages. You might see the URL of the search engine but not the URL of the web site. This happens with AskJeeve.com but the designer has thoughtfully supplied a 'remove frame' button (see Fig. 5.2). Press this and the target URL is revealed in the location bar. About.com is a useful

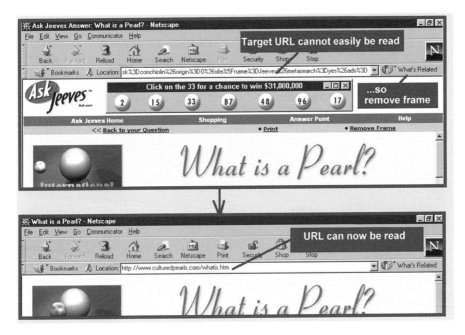

Figure 5.2   Removal of the frame to reveal the URL in Ask Jeeves

news/search engine which does not give you this facility. However, if you right-click with the mouse (as described in section 4.5) and select *View Frame Information* you will see the full web address of the target. Copy and paste this into your assignment as necessary.

6
Email

6.1 INTRODUCTION

There are many ways of sending written messages – by surface letter post, by airmail, by Fax, by pager and by email, among others.

Email is a convenient way of sending messages electronically. The message is typed so that it appears on a computer screen, a 'send' button is pressed, and after travelling via intermediate computers linked up by landline or satellite links, the message arrives at the computer of the target person (Fig. 6.1). Nowadays, the receiving computer and screen can be a portable phone. Ordinary paper-based postal services are now so slow by comparison with email that they are sometimes called 'snail mail'.

The email address usually includes three or four sections. For example: george@cant.ac.uk

This address says

- george – the *name of the target*
- @ – *at*
- cant – Canterbury – represents *the provider*; here it is the *name* of the institution (Canterbury Christ Church University College)
- ac – academic – the *type* of target (an academic institution, usually a university)
- uk – United Kingdom – geographic *location* (the country code). These are usually self-evident: uk = United Kingdom, fr = France, se = Sweden, de = Germany, be = Belgium, au = Australia, es = España (Spain).

The part after @ is the *domain name*. Across the world there are computers called *domain servers* which contain lists of domain names and their addresses. When an email is sent, the message is broken into packets

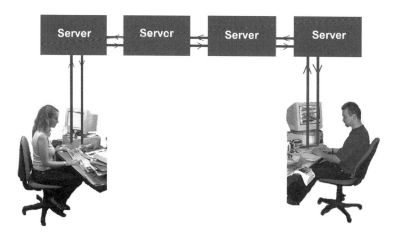

Figure 6.1 Computers called 'servers' link correspondents in email

which are sent separately across the internet. A copy of the destination address is attached to each packet. The packet passes to a *mail submission computer* which checks that the packet has somewhere to go. If the address is incorrect and the packet has no clear destination, the packet is bounced back to the originator with an error message. If the address is correct, other computers called *routers* organise a pathway for the packet to travel across the internet. When all the packets have arrived, the mail server at the destination reassembles the message and puts it into the *mailbox* of the recipient.

Because students are frequently peripatetic between their university and homes of their families, many students find it convenient to have their own email account, separate from that given to them by their university. Free email addresses are offered by Hotmail. It takes only minutes to get a free email account. Hotmail is hosted by Microsoft at:

http://lc3.law13.hotmail.passport.com/cgi-bin/login 4/4/01–22/7/01

6.2 ADDITIONAL USEFUL FEATURES OF EMAIL

Immediacy – email seems to have a psychological dimension which means that people respond much more quickly to emails than to paper post. This may relate to the fact that emails can get immediate responses. Some firms have stopped their personnel from opening their emails first thing in the morning because they then spend the whole day dealing reactively with incoming mail. If you are easily distracted, you might find this postponement a useful strategy to apply to your personal time-management.

Attachments – documents such as image files can be attached to an email message. This is rather like sending someone a letter envelope in which there is both the letter and a photograph.

Try to be sparing in sending attachments. They can take up a lot of time for the receiver. My personal strategy is to have a folder which I call AAAA. Because this is the first letter in the alphabet, it sits on the top of all my folders and is immediately obvious. When an email arrives with an attachment, I highlight the attachment and then save it to AAAA. I can then look at the attachment at my leisure, before transferring it to somewhere more appropriate in the directory of folders. Figure 6.2 shows how this is done with an email package called Simeon. The issues are general for any package.

Save sent mail – you can save all your sent mail in a special folder. This is useful when you may need to prove that you did something.

Read receipt – you can arrange email such that *when the recipient reads the message, an acknowledgement comes back to you. Sometimes people claim they never received a message but you can prove that they received, and opened it. This could be useful if you are submitting assignments to tutors via email.* Equally, the tutor might use it to prove that you had read their comments after an assignment had been marked.

Message stacking – during an exchange of emails, you can accumulate a dialogue by simply adding their last email message to the others which have already been exchanged. In an email dialogue, it is conventional to add your comments *on top* of the previous one, and *not to delete* the previous messages. In this way, correspondents are reminded of what has gone before.

6.3 CONTACTING EXPERTS

The speed and ease of email, plus its association with the web, mean that it is relatively easy to find and contact experts. There are two opposing views on this:

1 that students should find information from the literature and should not need to contact experts;

2 that an expert will be pleased to be contacted by genuinely enthusiastic students.

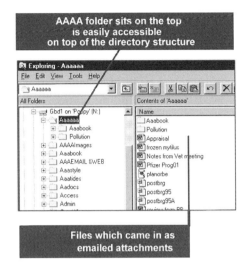

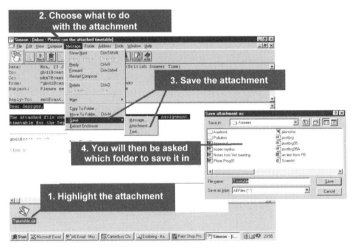

Figure 6.2   Use a folder called 'AAAA' to place your attachments until you have time to deal with them

If you are contacting an expert whose views are in the former category, you could get a brusque put-down. However, it is probably fair to say that most experts fall into category 2, as long as they are dealing with interested and motivated students.

It is important when contacting an expert *to be precise and knowledgeable about your question.* A pollution expert who receives questions such as '*I am doing a project on pollution and wonder if you can tell me what you know about it* quickly consigns such queries to the waste bin. The Executive Director of the Institute of Ecology and Environmental Management in the UK notes:

> ... We receive an increasing number of enquiries from students – either at university or preparing for A levels – for information relating to their projects. We try to help to the extent of our distinctly limited resources but there does seem to be an increasing trend for a student to tour the obvious websites, find a likely source, send some emails asking for help and then hope that the response will produce a finished project. I am afraid there have been some disappointed callers!
>
> (Thompson, 2001)

Only contact an expert if you have already done some research; contact with an expert represents the icing on the cake – not the cake itself.

Here is a transcript of an emailed dialogue between two people who are interested in the ancient art of *falconry*. Seeing this dialogue is rather like eavesdropping. It tells us about the subject and something about the people. I have quoted it extensively because it contains some interesting biology.

Ray is a novice who has just obtained his first hawk. Mike is an expert zoologist with many years experience of falconry. In this emailed dialogue, the messages stack up on the computer with the most recent message on the top. To help follow the developmental history of the dialogue, it has been inverted so that the first message is at the top. Note that there are misspellings and grammatical errors. This is normal in emails. It is a more immediate medium than snail mail, so most people take less care with the niceties of the language. However, *in my opinion, it is always wise to check through and remove the major errors*. Otherwise, you could appear to be careless!

6.4   RAPTORIAL BIRDS – A DIALOGUE BETWEEN ENTHUSIASTS

On Fri, 18 Aug 2000 11:54:14 +0100 Ray C. <ray@talk21.com> wrote:

>Dear Mike,
For quite a few years now I have harboured a great interest in birds of prey. I write to you knowing your

[1.] This is the first message from Ray to Mike. Ray had found out about Mike's interest and expertise in falconry by searching the internet

expertise in the field of raptors and would be gratful if you could answer some questions. I would be grateful if you could explain the difference between Harris Hakws and Goshawks. Is it merely the way these two specieshave evolved due to their geographic locations? I look forward to receiving your comments. >
> Ray

From: <mike@cant.ac.uk>
To: Ray <ray@talk21.com>
Sent: Friday, August 18, 2000 9:57 PM
Subject: Re: Hawks

2. This is the reply from Mike to Ray

>Dear Ray,
>Nice to hear from you and I will answer your question as best I can.
I think your question is why are Harris' different from goshawks ? Well gosses are Accipiters – 'true' or bird hawks – and Harris' are Parabuteos – literally 'like a Buteo' or like a buzzard. Genetically I believe that Harris' are nearer to buzzards than to accipiters. In South America the local Harrisses have shorter rounded wings and are woodland bird hunters. The Harrisses famous for catching large qround prey such as jack rabbits are the large Sonoran race from New Mexico. These sometimes hunt in family groups and tackle large hares in relays!
I hope that this is of help,
best wishes
>
Mike

>Mike,
>Thank you for your recent e-mail regarding the principal differences between Goshawks and Harris Hawks. As I mentioned before, my female Harris is currently moulting her feathers. Many people have asked me if this is the same for a wild raptor? Wild hawks fly daily to catch their food. Do they still moult their feathers like a captive bird?

3. This message comprises Ray's comments and further questions in response to Mike's reply

From: <mike@cant.ac.uk>
To: Ray@talk21.com>
Sent: Tuesday, September 26, 2000 5:10 PM
Subject: Re: Hawks

4. This message comprises Mike's comments in relation to Ray's second email

Dear Ray
Summer is a convenient time to allow a trained bird to moult. As you point out wild hawks are breeding and so need their feathers. Indeed female sparrowhawks moult their main wing and tail feathers while they are incubating and the male is feeding her. Large birds of prey like eagles may take several years to complete a full moult, just replacing a proportion each year.
Let me know when you get your Harris' going and we'll have a day out together.
Best wishes
>
Mike

Dear Mike,
>
I hope you and 'the bird' had a productive time in Scotland. I must admit that when we were up there I

5. The conversation continues but we can leave it here

wished every second that I had taken my hawk with me. Not to worry, maybe next time...! On a sporting note, I should be reclaiming my hawk over the next couple of weeks, especially when Kevin helps me put on her tail bell! Then we shall try and get a day out together.
Best regards
Ray

6.5 NEWSGROUPS AND LIST SERVERS

These are media by which *a number of people* can participate in a discussion at the same time, using basically the same methodologies as for email. With a newsgroup, participants place their messages on a virtual bulletin board which can be read by all the participants. With a list server, each participant receives a copy of any message which is sent to the list. For example, person A sends a message to the list server which passes the message to all the participants (Figure 6.3).

The systems are basically similar and it is necessary to go through a simple registration procedure in order to take part. This procedure will usually require you to offer and then confirm a password; your membership will be confirmed with you by email.

6.5.1   *Newsgroups*

Newsgroups (sometimes called *Usenet Discussion Groups*) are places on the net where there is open discussion. One person can raise issues with many others at the same time. For example, in Figure 6.3, person B *posts* (i.e. sends) an *article* (i.e. a message) to a newsgroup address. Anyone who has access to the address and who want to comment does the same thing. Their comments are visible to everyone concerned.

Newsgroups are classified into categories such as science-related, recreational news, social issues, controversial subjects, computing and miscellaneous. Newsgroups which are of interest to biologists often involve the extension *bionet* in their title.

To get access to a newsgroup first you have to search for one that suits you. You can usually do this through a service provider such as Yahoo at http://groups.yahoo.com 1/3/01–27/7/01.

When you have found your group, you will have to register and follow a simple sequence of steps to give you a *password* and *identifier*. The

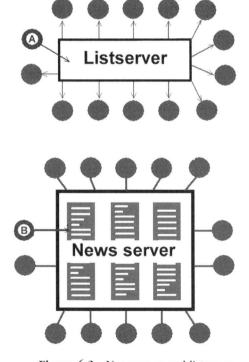

Figure 6.3   Newsgroup and listserver

newsgroup discussion will then be automatically sent to your email inbox and you can join the discussion whenever you like. A line of discussion is called a *thread*. When you join the discussion, you are 'following up a posting'. Some newsgroups have a *moderator* who volunteers to keep the discussion in line. Treat the discussion in the same way as you would a discussion on paper. Although there are simple rules to conducting yourself in a discussion group (Chellen, 2000), they can be summarised within the ideas of *being diplomatic and using common-sense rules of etiquette*. Bad etiquette can get you a bad name and messages can be saved and used against you in the future. For example, if you are corresponding with someone you want to impress (for example, someone who might at a later stage influence your career) it is not a good idea to post messages which contain spelling and grammatical errors.

In particular, you should beware of getting angry in the newsgroup (*flaming*) and saying things you might regret later. If you are going to be critical, make it positive, phrase your sentences diplomatically, use *emoticons* and sleep on the critical email before you send it. The commonest emoticons (look at these sideways) include:

:-) smile

:-(frown

%-) puzzled

:-I anger

;-) wink

:-P yah boo

6.5.2 Mailing Lists

For a mailing list on a list server, there are two addresses. One is the administrative address of the person or organisation that runs the list; this is the address you need to approach in order to *subscribe* (get on) or *unsubscribe* (get off) a list. By contrast, the list address is the one you use when posting information and carrying on the conversation. Anything you send to the list will be received by everyone on it. List server programs include Listserv, Mailbase, Listproc, Mailserve and Majordomo. Here is an example of the procedure for getting onto a list server.

Subject - Biomechanics

Step 1. A search engine takes me to http://www.topica.com/dir/?cid – 191 2/5/01–23/7/01 which gives a list of different discussion subjects (Fig. 6.4). I am told there are 542 lists in biology, and four in the subject area of biomechanics. I investigate. The four lists comprise:

- Bioinformatics - involves the San Diego bioinformatics community
- The Cleveland Biomechanics and Motor Control bulletin board
- Sports Biomechanics Forum
- Biomechanics and Movement Science list server.

The latter looks the most relevant to my interests.

Step 2. Topica asks me to subscribe, first asking for my email address and some other elementary information.

Step 3. I am careful to tick the box saying I DO NOT want to be told of any new products and services.

Step 4. A confirmatory email comes into my email inbox. To confirm that I am subscribing, I send a simple reply.

Step 5. I can now join the discussion.

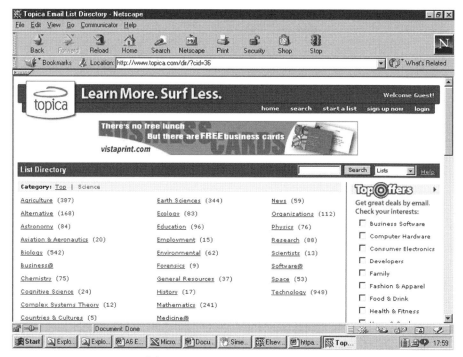

Figure 6.4 The initial Science list page at Topica

6.6 EXAMPLE OF A GROUP DISCUSSION – EXPOSURE TO IONISING RADIATION AND THE ONSET OF LEUKAEMIA

Usenet is an internet bulletin board covering a wide range of topics. Here is an example of a real discussion (the names have been changed) concerning radiation biology.

Sent: Friday, December 15, 2000 6:36 AM
Subject: Re: [srp] leukaemia & chronic exposure
Dear All,
I need to relate an exposure of 10 Rem (4 years ago) to induced leukaemia in a male adult. Please help regards from Nadia

Sent: Friday, December 15, 2000 10.30 AM
Dear Nadia
10 rem now more familiarly known as 100mSv risk factor for fatal cancer (leukaemia) according to ICRP60 is 4% per Sv so an exposure of 100 mSv gives a risk of 0.004
From Ian,
University Radiography technician

Sent: Monday, December 17, 2000 12.03 PM
To Nadia:
We often work on these cases in the U.S. The number given by Ian is the risk of contracting leukemia, but I don't believe this is the risk number you want. Since this person already has leukemia, you want to know the likelihood that his leukemia was caused by radiation. To estimate that, you take the ratio of the risk of contracting leukemia, the number below, to the total risk of contracting leukemia, the radiation risk plus the spontaneous risk. From Jenny, US Environmental Protection Agency

Sent: Monday, December 17, 2000 12.01 AM
To Nadia
The estimates of risk for radiation induced leukaemia in males according to BEIR V Report are quoted in Table B-9 of ICRP

Publication 60 (1990 Recommendations of the International Commission on Radiological Protection) as follows ...
Excess lifetime mortality after exposure to 0.1 Sv (i.e. 10 rems) acute uniform whole-body low LET radiation

Age at exposure : probability of death
-------------- : -----------------
```
        5 years  :   0.00111
       15 years  :   0.00109
       25 years  :   0.00036
       35 years  :   0.00062
       45 years  :   0.00108
       55 years  :   0.00166
       65 years  :   0.00191
       75 years  :   0.00165
       85 years  :   0.00096
       Average   :   0.00110
```

from John Smith, The Royal Hospital, UK

Thus members of the list have given Nadia some concrete support in solving her problem.

6.7   WORKING OFF-LINE

In order to get access to the articles in a discussion group, you may need to use a program called a '*newsreader*'. Since you may be paying charges for being on-line, it is more economical to download all the message in your newsgroup into a file on your computer and then investigate the discussion while you are off-line. Thus you pay the minimum in telephone charges. In Netscape, the newsreader is Netscape Collabra.

6.8   CAVEATS ABOUT EMAIL, NEWSGROUPS AND LISTS

Email, newsgroups and lists concern discussion in one form or another. This necessarily means that the information is not solid. It has not been peer-reviewed and is merely in the *realm of opinion*. While your tutor will be pleased to see that you are investigating these aspects of the medium, you will get little credit if you treat these sources of information as if they were equivalent to a journal article or a paragraph in a textbook. It is better to use these sources as a way of getting a 'feel' for what is happening in the discipline.

In such media, be particularly careful that you are not picking up information from people (e.g. other students) who are probably less well informed than you are. Remember that '*empty vessels sometimes make the most sound*'.

As you get further up the educational ladder, email, newsgroups and lists become increasingly important and are an essential part of the armoury of a research student. Any aspiring research student must start to *network* as early as possible in their research career. This means attending conferences, presenting posters, making oral presentations and making academic contacts. Emails, newsgroups and lists help the research student in this. When used properly, these systems promote a scientifically creative spirit and enhance motivation.

6.9   HAVING A PROFESSIONAL ATTITUDE

Because of the time which can be taken up with emails, firms are becoming wary of employees spending a lot of time in working hours on emails which are essentially personal. As with telephone calls, most employers will turn a blind eye as long as the system is not abused. This is also true of universities. However, it may be in your own interest to have an email address which you *reserve for personal business*. Hotmail seems to be a reliable system (see section 6.1). This could be worth doing because:

- it might be a good professional attitude to entrain for when you enter paid employment;

- in the university system, your emails can be monitored;

- when you leave university and change addresses, it might be convenient to have one, steady email address; your friends will always know how to contact you.

7
Approaches to Preparing a Biosciences Assignment

7.1 *INTRODUCTION*

This section concerns general advice for preparing an assignment. Although it is not specifically relevant to the internet, many students who lack skills in this area have their deficiencies magnified by the electronic media they are using. For example:

1 Because of the perception that the information is abundantly available and easy to obtain, many students leave their preparation to the last moment. This does not allow for technical problems such as software which crashes, printers which break down and dogs which chew the diskette on which the information was saved without back-up. *Et cetera*!

2 Some students are overwhelmed by the amount of information available to them and the cut-and-paste possibilities. Frequently, such a student obtains good information from the internet but then pastes it together in a thoughtless and incoherent way. This kind of assignment has been seen by the majority of experienced university teachers.

3 Because many of the media possibilities are new, students don't realise that there are, nevertheless, important conventions regarding aspects such as referencing. Here, the watchword should be, first, that the reader can trace a statement or a justification to its source and, secondly, that a standardised referencing system has been used. Although there is a wide range of systems, the Harvard system is becoming the 'industry standard' in biology.

7.2 *GETTING STARTED ON THE ASSIGNMENT*

There is a Chinese proverb which says '*The longest journey begins with the first step*'. An English proverb says '*Making a start is half the job*' and another says '*Procrastination is the thief of time*'. All of these proverbs say the same thing: if you are motivated and excited by your subject, it won't be too difficult to make a start but you still have to make an effort. You can also be certain that if you leave it late, the printer or some other part of the infrastructure will break down and you will miss the deadline. If you complete the assignment early, the printer will stay in magnificent working condition. In laboratory sciences, this is the well-known phenomenon of Sod's Law (section 1.2.2). You should believe in Sod's Law and take out insurance. One piece of insurance is to *start early enough*.

7.3 CONTEXTS AND CHANGE-OVERS

Everything that exists has two properties:

1 *context* – for example, a picture on a wall is set in its context by the frame;

2 *change* – changing points in social behaviour are ameliorated by politeness; for example, a conversation begins with 'Hello', takes place, and ends with 'Goodbye'.

In the same way, make sure your work:

- is set in an appropriate proper *context*, usually by using appropriate literature in the Introduction and Discussion;

- has a *beginning*, *middle and end*, with a logical flow from one part to the next.

Revisit Chapter 3 on web citation. If preparing something in a more formal context (for example, on paper), ensure that where you use someone else's idea or information, it is referenced, usually with the authors' name and date of publication in the text, and with the full reference given in alphabetical order at the end. Most scientific journals now use the *Harvard referencing system*. You should seek the advice of your tutors for what style they prefer but almost certainly in biosciences, you will be expected to use a Harvard-type method in your assignments. Refer to an appropriate journal to identify an exact style and get into the habit of being punctilious with the layout of your references. A scientist might spend three years working on a piece of research and send it to a journal for publication but ruin the whole thing by not getting the

4. References

References in the text should take one of the following forms: 'Nisbet (1973) said ....' or '.... (Nisbet, 1973) ....' or '.... (Merdsoy & Farley, 1973; Nisbet, 1973; Anderson, 1980) .....'. Avoid *loc cit*. The first time a paper with multiple authorship is referred to, give all the authors' names; thereafter use the form X *et al.* References are listed alphabetically; the styles of entry to be used are:

ANSELL, A.D. & TREVAILLION, A. 1970. Brood protection in the stenoglossan gastropod *Bullia melanoides* (Deshayes). *Journal of Natural History*, **4**: 369-374.

DUNCAN, C.J. 1975. Reproduction. In: *Pulmonates* (V. Fretter & J. F. Peake, eds), **1**: 309-365. Academic Press, London.

ELLIS, A.E. 1926. *British snails.* Clarendon Press, Oxford.

NISBET, R.H. 1973. The role of the buccal mass in the trochid. *Proceedings of the Malacological Society of London*, **40**: 435-468.

RUNNEGAR, B. & POJETA, J. 1985. Origin and diversification of the Mollusca. In: *The Mollusca*, **10**: *Evolution* (E. R. Trueman & M. R. Clarke, eds), 1-57. Academic Press, London.

THORSON, G. 1958. Parallel level bottom communities, their temperature adaptation, and their "balance" between predators and food animals. In: *Perspectives in marine biology* (A.A. Buzzati-Traverso, ed.), 67-86. University of California Press, Berkeley.

Journal titles are cited in full. Print the volume number in bold or underline in wavy. Repeat authors' names and journal titles in multiple entries and avoid *ibid*.

Figure 7.1 Reference instructions for the guide to authors from the *Journal of Molluscan Studies*

references right and following the publisher's style requirements for them. Similarly, the language should be clear and concise. It ought to be a point of honour amongst scientists that they write good English. Refer to Booth (1984) for excellent recommendations on scientific writing and presentation. Try to make use of the grammar and spellcheckers on your computer. At the very least they will help you to improve your control of grammar and spelling. However, be aware that there are errors, particularly in spellcheckers, so keep a good dictionary to hand. My favourite dictionary for general English, which includes a good range of scientific words, is *The Oxford Reference Dictionary* with illustrations, edited by J. M. Hawkins, published by Clarendon Press, Oxford.

Figure 7.1 shows the reference instructions from the Guide to Authors for the *Journal of Molluscan Studies*: http://www.sunderland.ac.uk/~es0mda/jms.htm#Journal 6/6/00–27/7/01.

If you are doing your assignment on a word processor, make sure that you have *back-up copies* at every stage.

Here is the *ultimate catastrophe*. You have almost completed a piece of work called 'Essay.doc'. You accidentally open a new file in which you write nothing, call it 'Essay.doc' and save onto the top of the previous file, obliterating all the contents for ever. If you have an autosave function operating, the computer may save the empty Essay.doc onto the previous Essay.doc by itself. You might retrieve the situation by using Windows Explorer or File Manager to find the folder in which are stored all the TEMP files. These are temporary files which are accumulated during an on-line session and may contain a previous version of Essay.doc. Usually, they contain everything except the file you want (Sod's Law).

7.4   *GOOD PRACTICE FOR AVOIDING CATASTROPHE*

Here is good practice which you should always adopt:

1 Save onto the hard drive *every 10 minutes* or so. You can do this automatically by File>Save as>Options>Automatic save every... and you can also use these menus to set up back-up saves.

2 Give each generation of the assignment a progressively *changing filename*. For example,

    Essay1.doc
      Essay2.doc
        Essay3.doc etc.

3 Save text files onto *floppy disc and paper*.

4 Never delete files. Save them into a *'junk' folder* until you are sure they are not needed.

In the interests of boosting morale, If the catastrophe has happened to you, here is a true story which might help.

In the nineteenth century, the philosopher and historian Thomas Carlyle (http://www.bartleby.com/65/ca/CarlyleT.html – 29/11/00–17/7/01) spent several years writing his massive work *The French Revolution* using a quill pen and ink. When it was finished, he went out to celebrate. He had, however, engaged a new maid who came into his room and, thinking the papers were a mess, burnt them all. Carlyle returned to a pile of smoking ashes. It took him two years to re-write the book and he always maintained that the re-written version was much better than the original. So now that Sod's Law has meant that your assignment is destroyed, you know what you have to do...

7.5 WRITING AN ESSAY

7.5.1 The Mediocre Approach

1 Look up the main issues in the contents lists of several books or web pages.

2 Index the appropriate pages – with Post-it bookmarks in a paper text, or as 'Favorites' or 'Bookmarks' for web pages.

3 Summarise each bookmarked item.

4 Re-organise these summaries into a logical order.

5 Write the essay, i.e. put the summaries together.

The problem with this approach is that although appropriate *things* are being done, little *thinking* is being done. If you do an essay like this, you are acting like a machine and will get mediocre marks.

7.5.2 The Efficient and Appropriate Approach

1 Visit the library and make notes from primary (research articles if the journals are available) or secondary sources (textbooks).

2 Conduct a computerised literature search in the academic literature and get a few key articles.

3 Use the information gained in (1) and (2) to give you background on the subject area. Use this background to start efficiently searching the internet.

4 Gather, collate and assimilate the information as notes.

5 Write the essay as if you were in an examination, allowing the mind to do the work by itself. Your brain is a tool and you should allow it to work; brains are good at thinking, so let your brain think. Write the essay using your own words. Remember that you are doing this to learn, and that you therefore need to assimilate.

6 The final stage is to return to the notes and make the essay concrete, with appropriate references and attributions, including web links.

7.5.3   The Zen Approach

Perhaps the best approach is bring some zen into your essay writing. Do items 1 and 2 in 7.5.2 and then read around the subject. Relax, have a cup of coffee, go for a walk and go into a meditative mode by trying to '*think without thinking*'. Allow your own creativity and imagination to play their role. When the time seems right, sit down and write an extended plan, without reference to the literature. Just allow the ideas and *theme* (the story line) to erupt, by themselves, from your own mind. Once you have written the theme go back to the literature and fill in the details. Make sure that you have evidence to support every part of your argument. The evidence should be concrete examples, such as the names of relevant species or processes. It was this more integrated approach, using all the capacities of the human mind, which made the phoenix of Thomas Carlyle's *French Revolution* a better book than the original.

7.5.4   A Cynical Approach

Put yourself into the shoes of your tutor. What do you expect of a student?

As a tutor, you hope that students understand and know the basic material you supply in class. You also hope that students have done enough extra reading to put their learning into a context. Although there are specific, exciting, appropriate things a tutor wants students to read, tutors usually and increasingly feel that the problem is not that students read the wrong things, but that they don't read anything at all, or at least, don't read anything unless they have to.

This sentiment probably relates to the fact that tutors are almost always old and crabby. Paraphrasing the words of James Taylor, the 1970s cult songwriter, their smiles have turned upside down. Most tutors seem to believe that when they were young, students read widely, got their assignments in on time, the sun was always shining and the snow was always white.

For the student, the problem now is that there is so much information that it is difficult to define 'reading widely'. You can't read everything, so what should you read?

A legitimate but cynical way to play the game is as follows. Suppose I am a student who has to do an assignment on (say) *endangered species*. Using the cynical approach, I can find an appropriate web site (or textbook) and poke around in it until I have found only one or two relevant facts. These facts can then be woven into the assignment and referenced appropriately.

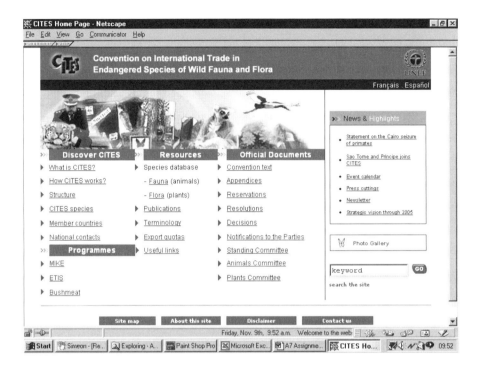

Figure 7.2 Web page of CITES

For example (all sites visited on several occasions between 15/11/00 and 27/7/01):

>http://www.unep.ch (i.e. the United Nations Environment Programme)
>http://www.cites.org (i.e. the convention on trade in endangered species – CITES)
>http://www.cites.org/eng/site_map.shtml (i.e. the CITES English-language search engine)

I select CITES databases > fauna (Figure 7.2).

My eye is caught by 'specific reservations' which sounds like participants who don't fully agree with the sentiments of the CITES protocol. Intrigued, I scan down the list of species and find that the UK has reservations about conservation of *Mustela erminea*. I don't know what this is, so I go to a general search engine and find this site:

>http://www.bio.bris.ac.uk/research/mammal/weasel.htm

This academically based site tells me that the *Mustela erminea* is a stoat. It seems a shame that there should be reservations about conserving this creature. The site also gives a nice image I might be able to use in my assignment. Now all I have to do is to weave this information into my assignment, thereby persuading the tutor that I've done more than the minimum and have done some 'extra reading'. For all the tutor knows, I have read a chapter of a book on mustelids. Although it's a cheat, I have nevertheless, extended my knowledge. For example, I now know something about UNEP and CITES, I know what a stoat is, and I know that the UK has reservations about the protection of these mustelids.

So, you can play the game cynically, appease your tutor and *still* extend your mind. Obviously this is a strategy of last resort. You should really be following the recommendations of section 7.5.3.

7.6   *WRITING A PRACTICAL REPORT*

All practical reports, at all levels, follow the same line. The writer describes what they set out to do (the *objectives*), describes what they did (the *protocols* or procedures) and what was found out (the *outcomes*). School pupils are encouraged to describe their practical work in exactly this way, but new undergraduates are often confused by the apparent conflict between what happened in school or college and what is expected of them at university. The confusion arises because of the semantic difference between the verb *to conclude* and the noun *conclusion*. In relation to the former, school pupils are asked what they conclude from an activity, and there will then be a class discussion of what it all means. However, the noun means the *end* of something. For a university bioscience student, the conclusion should be the *conceptual finale*; except for the references and possibly an Appendix, there should be nothing substantial after the conclusion.

All practical reports, including theses and dissertations should be laid out in this order:

- *Title*
- *Introduction* – setting the context and referring to the literature
- leading to the *Aims*, or objectives – clearly stating or inferring the hypotheses to be tested
- *Materials and Methods* – describing what was done

- *Results* – describing what was found out

- *Discussion* – what the results mean, in themselves and in the context of the literature

- *Conclusion* – briefly addresses the questions 'So what?' and 'What now?'

- *References* – laid out very carefully in the correct style

- *Appendix* or several Appendices

This method of presentation is traditional. Although it is a clear way to describe an experimental investigation, it is, however, a myth that scientific work follows this kind of path. Every biologist should read *The Art of the Soluble* by the Nobel prize-winner Peter Medawar to find out the true methodology which underlies research in the biosciences:

http://www. bartleby.com/65/me/Medawar.html 15/11/00–27/7/01

The *Materials and Methods* section should not be written in the 'recipe style of the imperative'. In other words, you should not write a recipe which tells people what to do (imperative – '*Pour the water in the beaker*'), because they probably won't do it anyway. You should say what you did (simple past tense – '*The water was poured into the beaker*'). There is an increasing and welcome trend to make the Methods more personal (first person singular – '*I poured the water into the beaker*'). However, too much use of the personal pronoun ('I' or 'we') can irritate the reader, so find ways of keeping it under control (not '*Me and my friend Dwayne spilt the soup all over Dr Jones and then went with Mandy to look for limpets*').

The *Results* section should give a *prose description* of what the results *are*, plus summary tables and diagrams. Where a statistical result is quoted, supply <the statistical value>, <the degrees of freedom> and the <probability of the null hypothesis being correct>, e.g.

$t = 7.2$ d.f. $= 56$ $P<0.01$
$r = 0.87$ d.f. $= 120$ $P<0.05$
$F_{1,9}$ ~2.5 $P>0.05$

These statistics should run along with a statement of the result:
e.g. The two sets of tissues consumed oxygen at significantly different rates ($t = 7.2$ d.f. $= 56$ $P<0.01$).

The *Discussion* section is one of the places where your imagination comes into play. The Discussion should address the reasons for the results being as they are. For example, the question of *why* two tissues might respire at different rates will be addressed, in the light of the literature.

The *Conclusion* summarises the relevance of the experiment and explains what might be the next steps.

More detailed materials such as tables of raw data are put in the Appendix.

Monographs, theses and commercial reports usually benefit from having a hierarchically structured numbering system. For example:

1 Introduction
    1.1   Inception of the project
    1.2   Immediately relevant literature
    1.3   Contextual literature
    1.4   General aims
    1.5   Specific objectives

2 Materials and methods
    2.1   Site locations
        2.1.1 General location
        2.1.2 Site maps
        2.1.3 Site descriptions
    2.2 Sampling methods
        2.2.1 Activities on site etc.

7.7 *BIBLIOGRAPHIC REFERENCING SYSTEMS*

Bibliographic software packages such as *Procite* are becoming increasingly useful. If, during your undergraduate career, you intend to go into research, it would be worth getting started with one of these packages quite early. Figure 7.3 shows a typical entry stored by such a program.

For example, it is possible to organise an automatic keyword search of research databases on a system such as Web of Science (part of which used to be called 'BIDS'). New, relevant literature can then be *automatically downloaded* into your email inbox. The bibliographic software picks up these references and organises them into your personal database. Using such a program, the reference list can be formatted for a particular tutor (or journal), and then reformatted for another tutor (or journal) as necessary. Figure 7.4 shows a short reference list adapted first for the journal *Biochemistry* and secondly for the journal *Nature*.

A typical range of styles offered by a bibliographic program (Procite) includes the following:

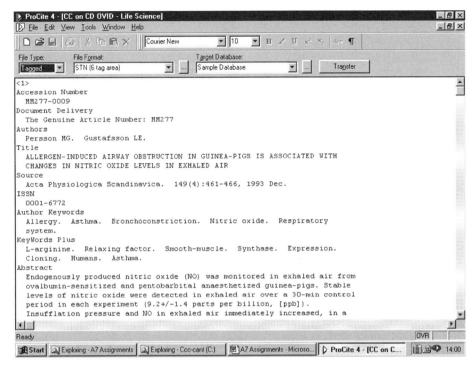

Figure 7.3 A typical bibliographic reference program document

American Chemical Society
American Medical Association
Biochemistry
Biochimica et Biophysica Acta
Index Medicus
Journal of the Chemical Society
Journal of Chromatography
Journal of Physics
Modern Language Association
Nature
Physical Review
Science
USGS-US Geological survey

Some universities provide information regarding support for students who are not sure what systems they should be using. For example, comprehensive and comprehensible advice can be obtained at:

http://www.canterbury.ac.uk/list/organise/or_s2.htm 3/6/01–27/7/01

For the journal _Biochemistry_...

Aroney, Manuel J.; Davies, Murray S.; Hambley, Trevor W., and Pierens, Raymond K. (Department of Inorganic Chemistry, The University of Sydney, Sydney, NSW 2006, Australia). A study of the polarities, anisotropic polarisabilities and carbonyl infrared vibrational frequencies of the complexes $[M(CO)_5\{P(OCH_2)_3CMe\}]$ (M = Cr, Mo or W) and the crystal structure of $M(CO)_5\{P(OCH_2)_3CMe\}]$: Evidence for p-acceptor behaviour in co-ordinated phosphorus. Journal of the Chemical Society, Dalton Transactions: Inorganic Chemistry. 1994; 91–96.

Chum, H. L. and Baizer, M. M. The electrochemistry of biomass and derived materials. Washington, D.C.: American Chemical Society, 1985. (ACS Monograph; 183).

Gilman, Alfred G.; Rall, Theodore W.; Nies, Alan S., and Taylor, Palmer. The pharmacological basis of therapeutics. 8th ed. New York: Pergamon; 1990. 1811 pages.

Harred, John F.; Knight, Allan R. and McIntyre, John S., inventors. Epoxidation process. Dow Chemical Company, assignee. US Patent 3,654,317. 1972 Apr 4.

For the journal _Nature_...

Aroney, M. J., Davies, M. S., Hambley, T. W. & Pierens, R. K. _Journal of the Chemical Society, Dalton Transactions: Inorganic Chemistry_ 91–96 (1994).

Chum, H. L. & Baizer, M. M. _The Electrochemistry of Biomass and Derived Materials_ (American Chemical Society, Washington, D.C., 1985).

Gilman, A. G., Rall, T. W., Nies, A. S. & Taylor, P. _The Pharmacological Basis of Therapeutics_ 8th edn (Pergamon, New York, 1990).

Harred, J. F., Knight, A. R. & McIntyre, J. S. US Patent 3,654,317 1972).

Figure 7.4   The same reference list, organised for two different journals

7.8   MAKING AN ORAL PRESENTATION

7.8.1   Introduction

The strongest advice is:

> _Talk TO your audience. Do not set up your presentation so that you are reading something AT the audience._

The four most common ways of supporting an oral presentation are

- _slide_ presentation,

- _overhead transparency (OHT)_ presentation,

- _Powerpoint_ presentation,

- _html based_ presentation.

The human mind has a problem listening to material which is being read aloud. It is usually too dense and difficult to follow and you will lose your audience within three minutes. It is far better to talk to your audience than to talk at them. Try to make eye contact with them, read their body language and pace your presentation in response. At a recent EU scientific conference, a few delegates read their presentations without a single visual aid. This seemed arrogant and the audience soon stopped listening.

7.8.2 Preparation of a Story Line

Decide what your theme or story line will be and make sure it has a beginning, middle and end. I find it useful to tear a sheet of A4 paper into four equal pieces and using a felt tip pen (because the tip is relatively thick – it limits the amount of detail you can insert), write on each piece of paper what will go on each slide or OHT. These can then be shuffled into a sensible *line of argument*. On average, expect to get through one slide, Powerpoint screen or OHT in two minutes.

Organise the text on each slide or OHT with *bullet points*, which you can address in a relatively informal way. The bullet points will remind you what you want to say. Thus you don't really need to have other supporting material.

7.8.3 Preparation of the Materials

The major methods are:

1 via word processor and then print or photocopy onto an OHT, or photograph onto a diapositive slide,

2 via Powerpoint, present orally while online and/or use OHT (or expensively made diapositive slides),

3 via .html to make a web page, present orally while online and/or print onto an OHT (or expensively made diapositive slides).

Powerpoint presentations are effective but, as for web pages, you should not use all the gimmicks, as this can irritate an audience. The tables are also troublesome, and older versions will only use enormous bitmap (.bmp) images. This makes it difficult to save the presentation onto a floppy disc for carrying from one venue to another. Powerpoint presentations have a recognisable style. You might consider trying to do something more creative to make your presentation different from the norm. However, remember that current data projectors are not powerful enough to give a clear presentation in a relatively light room and the audience often ends up straining to see detail on the screen. Colour contrasts which look good on a computer screen often look terrible when projected in a large room.

Hyperlinks in an .html presentation can be useful but don't jump around too much between web pages or your audience will not be able to follow you. I am now preparing all my normal lecture material as web pages for students to download.

Although it is technically feasible to download presentations from your own server as you deliver your talk, it is totally unreliable at present. Assume Sod's Law will operate and *always* have back-up (usually OHTs) for any kind of electronic presentation. As a young university teacher, I decided to base the whole of my first, two-hour teaching session on a mammoth slide presentation. The projector caught fire at the second slide and I had to *ad lib* the rest. It was awful. Now, whether I am going to make a Powerpoint or a slide presentation, I always travel with a back-up of my conference talk on OHTs. If the worst comes to the worst, they can always be held up to a window and still be visible.

7.8.4   *Oral Presentation Checklist*

Always:

- remember you are primarily giving your oral presentation for the benefit of the audience, *not yourself*,

- begin by saying what you will tell the audience, tell the audience what you want to tell them and, at the end, tell them what you have told them (i.e. have a beginning, middle and end, with contexts),

- have visual images in your presentation,

- make sure the text is *large enough*,

- make a *trial projection* of your OHTs or slides in an auditorium to see whether they are readable at the back. You will be surprised at how big the text should be,

- *ask a friend* to listen to your talk. Don't try to cover too much. You will be surprised at how little time you have,

- *leave time for questions*. A question from the audience is the highest compliment they can give you. It is more valuable than applause because it shows that someone stayed awake. An audience will give you applause out of *politeness*. They will ask you a question because they cared enough to listen to what you had to say.

Some of the advice given above is also invaluable for job interviews.

7.8.5 Preparing a Poster

Posters are increasingly important methods of presentation, particularly at conferences. One of the most impressive methods of preparing a poster using electronic media is to prepare the images in an image-handling program such as Coreldraw and then paste about 8–12 images onto a single Powerpoint slide, with appropriate linking text. The single slide comprises a single sheet of poster which can be produced on a large glossy paper by a printing firm or the university reprographics unit. Figure 7.5 shows a typical poster produced in English at a French conference.

However, such posters can be difficult to carry undamaged on public transport. Where I have a complicated journey by public transport, I prefer to compose a poster in Word by using several A4 sheets of images and text, with linking card arrows. When I arrive at the conference centre, I place these sheets in a logical sequence on the poster board with Blu-Tack or sticky-backed Velcro.

No matter how the poster is prepared, a successful impact depends on some key issues:

- The poster should be *aesthetically pleasing* with appropriate combinations of coloured images and text.

- The main text should be readable at a distance of four metres.

- The poster should have a clear *title*, *contact address* and means of contacting you at the conference.

- There should be an easily recognisable *flow* to the argument of the poster.

- *Less is more*. Don't pack the poster with too much information. (This is also true of oral presentations. You don't have to tell the audience everything you did. Just present enough to titillate them.)

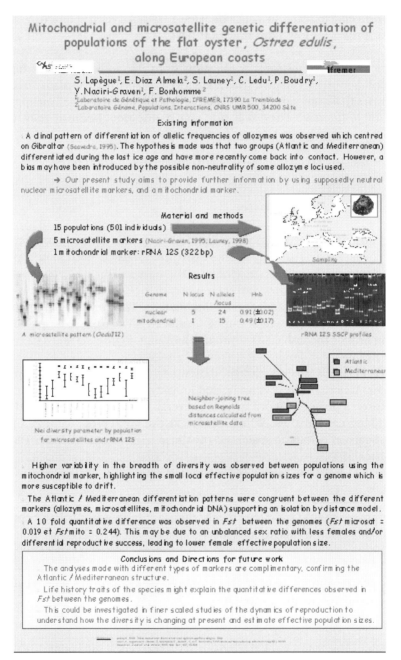

Figure 7.5 Powerpoint poster by French researchers, dealing with the genetics of the flat oyster

7.8.6 Finally...

For all presentations, try to come out of your own shoes, and *put yourself into the shoes of your audience*. What do they want to read or hear? What can they cope with? Can you entertain as well as inform?

Obtain a copy of Booth (1984) and read it from cover to cover. It is only a small book but if you follow the advice therein, it will make an enormous difference to your career no matter what you do in the future. Barrass (1990) and Dussart (1990) might also be useful.

8
Pinching Materials from the Web

8.1   *INTRODUCTION*

Pinch . . . 'If someone pinches something, they steal it; an informal use. *I pinched four pence from the box.*' *BBC English Dictionary* (1992)

The example in the above definition implies that pinching is less serious than stealing and is relatively trivial. Let us presume that 'pinching' is the web equivalent of a *little white lie*, i.e. something that isn't quite correct but something which occasionally has a *pragmatic* usefulness.

Please ensure that you have read and subscribe to the strictures in the previous, important, sections on *integrity* (1.1) and *plagiarism* (3.7). Let us suppose that you nevertheless want to copy some things from the net for your personal use. In effect, you want to pinch it.

There are four main things you might want to pinch from a web site:

1   *Images*

2   *Bits of text or html code* on which a web site is based

3   Whole *web pages*

4   Complete *web sites*

You might also want to poke around the structure of a page to see what it is made of, and see if there is anything worth pinching.

8.2   *TO PINCH AN IMAGE . . .*

Use a browser to find the web page you are looking for. In this case the page is at:

http://www.cant.ac.uk/depts/acad/science/staff/Paintings/Gallery.htm

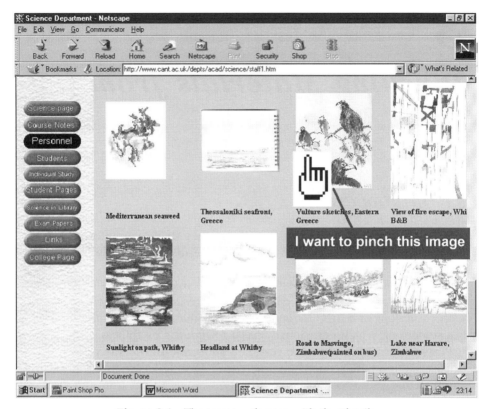

Figure 8.1   The target web page with thumbnails

Images on web pages are frequently presented as *thumbnails*. Suppose you wanted to obtain the full image associated with vultures in Fig. 8.1.

Move the cursor over the target thumbnail and if the thumbnail is hyperlinked, the cursor arrow will change into a *pointing hand*.

You will now be taken to a full-sized version of the image or whatever the thumbnail was linked to. Carefully compare the thumbnail and the image in Fig. 8.2 to spot the differences.

To pinch the large image, place the cursor anywhere on the image and press the RIGHT-HAND mouse button. This will produce a *drop-down menu* (Fig. 8.3). Select *Save Image As* in this menu and highlight it with a single click on the left mouse button.

You are now given a drop-down menu which shows your *folder structure*. You can decide into which file you want to save the image (Fig. 8.4).

Press the *Save* button and the image is now in your possession . . . but remember that it does not belong to you. You have pinched it. If you want

Figure 8.2   The full-sized image

to do anything with the image, such as use it in an article, you should get permission from the author. You can contact the author by email. It is polite to ask. However, if several people in your class might want to use the same image, perhaps someone (such as your tutor) can ask on behalf of the class.

8.3   TO PINCH SOME HTML CODE...

Remember that everything you see on a web page is supported by the *hidden* html code (see Introduction).

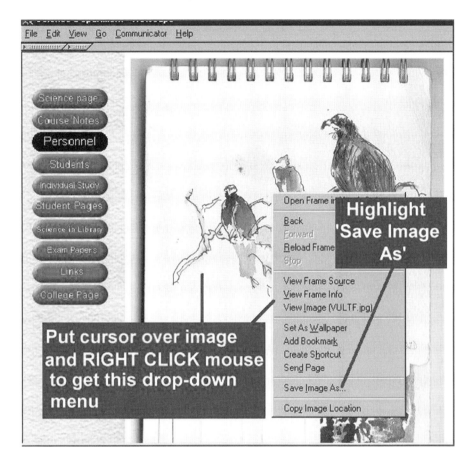

Figure 8.3   Finally deciding on the image you want to pinch

For example, on this page (Fig. 8.5), there is a counter which counts the number of times someone has visited the web page, and displays this count as a number of *hits*. This particular counter, produced by a firm called LinkExchange, is called 'Fast Counter'. It is available for free from Microsoft bCentral at:

http://more.bcentral.com/fastcounter 1/12/99–27/7/01

Suppose I want to pinch the code which leads to this counter, possibly to put it somewhere else on the page or even on another web page altogether... in other words, move the counter.

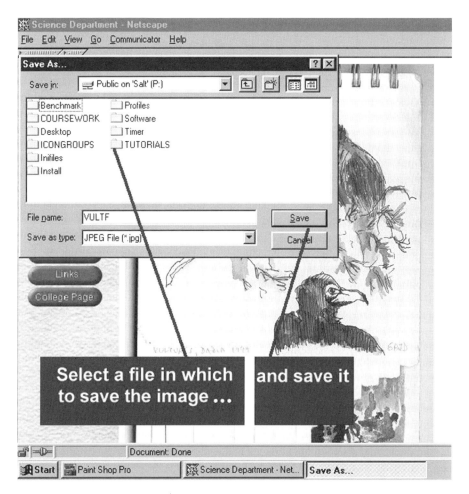

Figure 8.4   Saving the image into your folder

The cursor is moved over the object in question (the image of the counter).

Click the RIGHT-HAND mouse button and, as before, this reveals a drop-down menu (Fig. 8.6). On this occasion, what is wanted is the html source code.

Highlight *View Source* in the drop-down menu.

The underlying *html code* is now revealed. If you understand the basic commands, as described in Figure I.12 and in texts such as Crumlish (1999), it is relatively easy to find the piece of code to be pinched (Fig. 8.7).

In this case, we look for the words *fastcounter.linkexchange.com*.

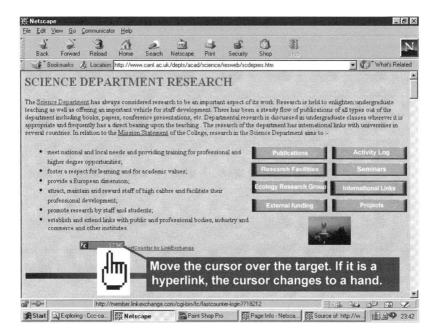

Figure 8.5 Moving the cursor over the target object for which the code is required

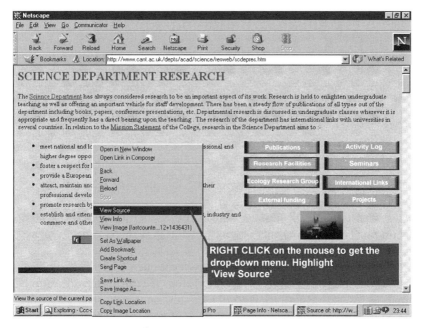

Figure 8.6   Getting the drop-down menu again

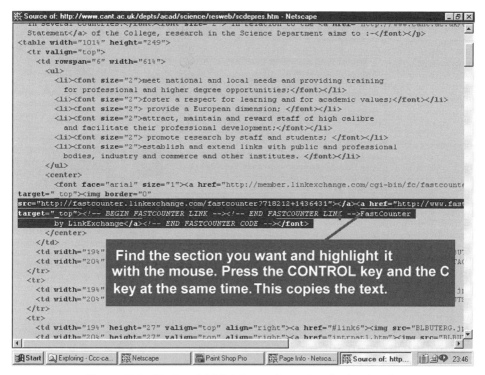

Figure 8.7   Finding, highlighting and copying the html code

Once found, the relevant lines are highlighted. Now you can use a very useful trick for copying things without using the mouse. *Copy by using the Control button on the keyboard and at the same time holding down the C key.*

The code can then be pasted into another part of the program, or into another program. Note that in actual fact, we are not pinching the counter. The counter stays with Microsoft bCentral. What is being copied is the *hypertext link to the counter.*

8.4   PINCHING WEB PAGES

Having found your target page, click on *File* and then highlight *Save As* in the drop-down menu which is revealed (Fig. 8.8). You can then save into your folder as in Fig. 8.4.

Where you have referred to a web page in an assignment, you should *save a copy* of that page. If the tutor raises any queries, you will still have the evidence, even if the site goes off-line at a later date.

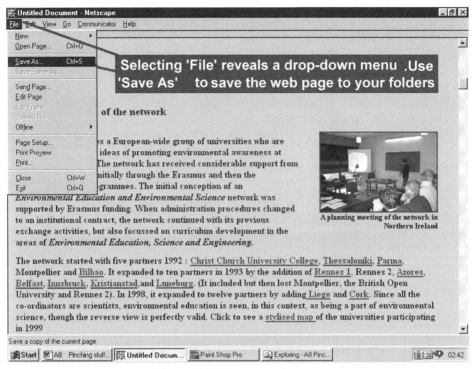

Figure 8.8   Saving a web page

If you are saving it *definitely and only* for your own purposes, there should be no copyright problem. However, be careful that you don't inadvertently copy text straight into an assignment. Learn to *précis*.

8.5   *PINCHING A COMPLETE WEB SITE*

There are programs which will allow you to copy a *complete web site* with all its links to your computer. Obviously this is drastic, since it could jam up the computer with hundreds of pages of web information. Only use such a process if you have good cause and know what you are doing. The circumstances in which one might want to do this are if a web site contained many pages of useful material that you need to look through at your leisure. Because of the call charges, looking through the whole web site on your own computer could be much quicker than downloading page by page across the internet.

The relevant software can be bought from:

http://www.tenmax.com/teleport/pro/home.htm 3/3/00–27/7/01

Figure 8.9 Selecting the target frame

8.6 POKING AROUND A WEB SITE

Web pages are often organised into *frames*.

Frames are like large cells in a table. When scrolling up and down one cell, the other cells remain stationary. This is useful when there are *generic links* that need to be available all the time the visitor is looking at the web page. For example, it is good practice for a web designer to construct a page so that there is always a 'home' button in view.

Let us assume you have come to a web page and want to find out a little more about what lies beneath it. The frames in a web page are not always obvious. You have to get a feel for the *rectangularity* of the way the page has been set up. Luckily, in the following way, it is possible to see whether your intuition is correct.

In Fig. 8.9, there are actually *two* major frames. The one on the left contains a set of red buttons and the one on the right contains the information of interest. The cursor is positioned in an appropriate place in

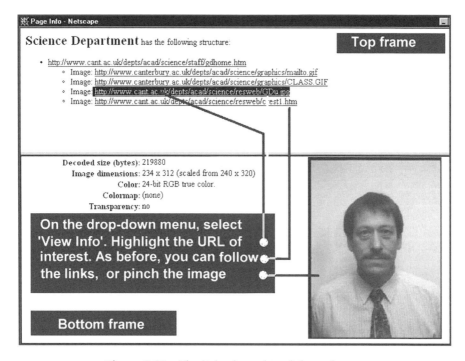

Figure 8.10   The links that subtend the web page

the frame and then the RIGHT mouse button is clicked, revealing the usual drop-down menu.

In this case, since we want information, *View Info* is highlighted and, after a double LEFT mouse click, reveals the horizontally split screen shown in Fig. 8.10. Unlike in Fig. 8.9, where the frames are side by side, the frames here sit one above the other.

The URLs underlying the site are shown in the top frame. Clicking on these will reveal further links which might be worth following up, or pinchable images which are shown in the bottom frame.

Finally, be careful to keep track of your saved images, pages and sites, and don't accidentally use them in a way which *breaches copyright*. In other words, make sure you don't pass them on to anyone, and remember you are not allowed to make any financial profit out of them.

While you are undoubtedly an honest person, *ignorance is not a defence under the law*. You could be sued in a court of law.

9

Web Sites which Present the Work of Students

When searching, be aware that where information is supplied officially or by *bona fide* employees such as research project teams, or lecturers, it is probably reliable. However, be wary. *Any* student can post information on a university web site. You might think it is 'official' when, in fact, the writer may know less about the subject than you do!

The Colorado State University in the USA has posted some work on the internet from students studying *chemical ecology*. These papers are generally excellent. Other students could critically evaluate these essays in relation to their own work. The university should be congratulated for being brave enough to put this work on-line. However, it is not immediately clear that this is the work of students. Although the file is headed 'Student Review articles', are these reviews the students have written or are they articles the students should review? In order to get the answer, I had to go backwards through the file structure as follows (5/12/00–27/7/01):

> http://www.colostate.edu/Depts/Entomology/courses/en570/papers_1994/papers.html
>
> http://www.colostate.edu/Depts/Entomology/courses/en570/papers_1994/
>
> http://www.colostate.edu/Depts/Entomology/courses/en570/
>
> http://www.colostate.edu/Depts/Entomology/courses/

Other places which present students' own writing includes universities at the following URLs:

http://gladstone.uoregon.edu/ *This site holds information from students working at the University of Oregon in the USA and is one of many that illustrates that student work is available on the net. Students should thus be aware that not all information held by universities has been produced by academics. 11/3/01*

http://www.micro.utexas.edu/courses/mcmurry/spring98 *Web-based projects by students at the University of Texas. Good but not perfect, and interesting to read. 26/3/01*

http://library.thinkquest.org/19037/therapy_links.html *This is a superb example of what students can achieve. Concerned with molecular genetics, the site is clear, concise, aesthetically pleasing and well-linked. It is part of the ThinkQuest site, in which students, with guidance, prepare scientific materials and put them on the web. Tutors should check the ThinkQuest site to make sure their own students are not plagiarising it. 30/3/01*

http://www.biology.washington.edu/fingerprint/dnaintro.html *Student web presentation on DNA fingerprinting. This is a good-quality performance, with accurate information. They make clear that they are students, which deserves a gold star. 30/3/01*

http://www.cs.tufts.edu/~cabotsch/bulloughs/protists/paramecium.html *This web page on Paramecium was signposted by a reliable scientific search engine ('About science'). However, it is an example of a piece of student work which has some failings but might easily be quoted by other students in error. Let's assume you are a student reading this; look at this web page with me. How do you know that I, as a university teacher, will be frowning? 30/3/01*

http://www.ccgs.nsw.edu.au/students/MarineTech/Mr%20Knox% 20Yr%209%20Web%20Marine/Star/Pg3.html *These Australian students have produced a jolly page with animations about starfish. Sweet. 30/3/01*

So the moral of the story is that we, the teachers, should encourage our students to post their materials on the web. As open source material, it helps everyone to be able to see what other people are doing, including other students. However, we must make it clear that this is the work of students and not professionals.

Part 2
Examples of searches

In this part, searches are undertaken in a number of fields which could be relevant to an A-level student or undergraduate in biosciences. The intention is to show the *thinking behind the search* rather than to provide a route which can be followed on-line. Indeed it is most likely that, given the transitory nature of web sites and pages, these sites will have changed by the time the reader chooses to investigate them.

The investigations include:

- investigating *microbes*;

- investigating a *university* web site;

- a typical general search, investigating the *wreck of an oil tanker*, the *Sea Empress*, which caused environmental damage when it went aground in Wales;

- a *government web site*, investigated within the context of the BSE (mad cow disease) crisis;

- an *on-line shopping trip* for a model of a dinosaur skull at a North American museum;

- a complex search which arises from a starting point which is predicated on the idea of independent thought in the enquirer. It uses the subject of *whaling* to investigate biases in the official and unofficial information which is available on the internet. The outcome is that the enquirer should not accept information at first sight;

- searching the pages of a *research group* (cancer research).

10
Microbes

In the following example, all sites were visited several times between 5/11/00 and 27/7/01.

Let us assume an assignment has been set which specifies a discussion of the relationship between virology, *BSE* (bovine spongiform encephalopathy) and *CJD* (Creutzfeldt–Jakob disease).

In *Step 1*, a search on Google with the keyword *Virology* leads to a site which calls itself 'All the virology on the www':

http://virology.net/garryfavweb.html

This grandiose title invites scepticism, but perhaps it's worth following up. Despite its title, the site has an interesting presentation and appears to have some useful links.

A quick look over the site reveals a link to *on-line virology course notes*.

Figure 10.1   Step 1

Figure 10.2 Step 2

Perhaps someone somewhere has already written the assignment? This site is starting to look useful and to justify its claims.

Step 2 is to:

http://www.virology.net/courseware.html

which, in *Step 3* jumps to another page on this site where there are lists of *courseware, including on-line tutorials* at:

http://www.tulane.edu/~dmsander/WWW/MBChB/6a.html

Some links on this page lead to course notes of staff at Tulane University in the USA. Curious, I follow this *sidetrack* to:

http://www.tulane.edu/~dmsander/WWW/MBChB/2a.html

and see a neat page on cells of the *immune system*, with illustrations.

Still sidetracking, I now wander off to:

http://www.tulane.edu/~dmsander/WWW/MBChB/6a.html

where there is an excellent page on *fungal pathogens*, with plenty of detail and illustrations, produced by Dr S. Kilvington at the University of Leicester, U.K.

Two things become clear. First, parasitic disease looks like an interesting subject. Secondly, academic staff deserve medals for not hiding such useful information away behind passwords. At some point, I'll return to this site.

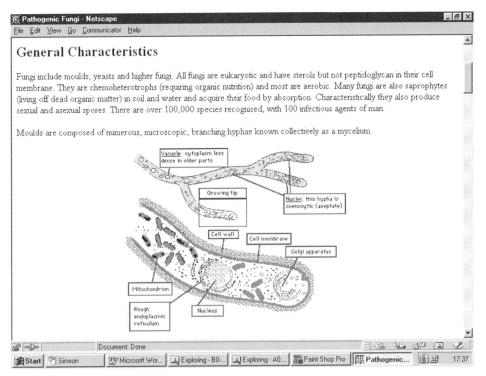

Figure 10.3 Step 3 (diversion to the University of Leicester)

The website name 'All the virology on the www is' beginning to look quite reasonable. In *Step 4*, returning to the on-line tutorial page at:

http://www.virology.net/Tutorials.html

my attention is caught by '*ATV's own tutorials*'. Without a clue on what this means, I follow the link.

The web page offers intriguing links such as '*A Dose of the Pox*', '*The Panama Puzzle*', '*Imagine you are a virus*' and '*How Now Mad Cow*'. These look *SO* interesting that the site gets bookmarked (or 'favorited' – see section 2.7).

'*How Now Mad Cow*' seems to be a relevant subject, so I follow this up by taking *Step 5*, which leads to a page dominated by the symbol for the British House of Commons at:

http://www.virology.net/Tutorials/cow/cow1.html

We seem to have slipped into politics! The web page describes how a

Tutorials:

- ATV's own Tutorials
- ATV's own Course Notes

Related Resources:

- Links to other Education Resources
- Virology Video Library
- Big Picture Book of Viruses
- Virology Bookshop

ATV Home

ATV Table of Contents

Submit a New Site

Online Tutorials

Click on the links below to get to the following interactive online tutorials:

- How Now Mad Cow
  How much do you really know about BSE? Can you get brain disease from eating beef? Decide for yourself...

  "A particularly lucid account of the current BSE controversy" University of Liverpool School of Biological Sciences.

- A dose of the pox...

  "An interactive tutorial that steps you through the discovery and ultimate eradication of smallpox using a series of questions. At each step, information and images are provided to help broaden ones appreciation of this scientific feat! A great teaching tool...and a lot of fun, too." Access Excellence.

- The Panama Puzzle
  Be present as history is made!

  "It's June 25th, 1900. Walter Reed needs your help to put an end to "yellow jack," a disease causing his men to drop like flies." Yahoo

Figure 10.4   Step 4

"There is no conceivable risk of BSE being transmitted from cows to people."

The Rt. Hon. Stephen Dorrell,
Minister of State for Health
Her Majesty's Government
3rd December 1995.

(click on the arrow)

© AJC 1997.

Figure 10.5   Step 5

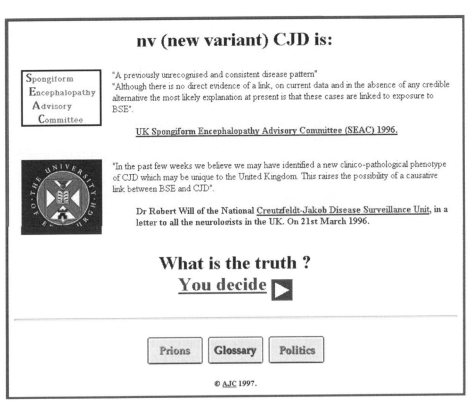

Figure 10.6   Step 6

politician in power in government at the time made a statement in 1995 including assurances that BSE presents no risk. (However, we now know that by 2001, over 80 people are dying each year of the associated disease Creutzfeldt-Jakob disease, compared with less than 10 before the BSE crisis.) The number of fatalities is rising. The web page challenges us to go on . . . and *Step 6* takes up the challenge by following the link to:

http://www.virology.net/Tutorials/cow/cow2.html

This site refers to a university and the Spongiform Encephalopathy Advisory Committee which sounds official and invites the reader to consider the validity of some political statements. This all seems to be getting rather contentious and before going further, I'd like to confirm the status of the person who wrote these pages. A diversionary step via 'AJC' at the bottom of the web page leads to the homepage of Dr Alan Cann of the

New Guinea, 1949:

Ibuso pushed the twig up through the small hole at the base of the skull, then moved it in a circular motion around the inside of the man's cranium. A thin grey paste began to drip from the skull into the bowl. It had been easy to remove the Shaman's head from his neck using the broad steel knife. The last time Ibuso had performed the funeral rights had been on her own husband, just under a year ago. Then, she had still used the stone tools copied from those which her family had used for centuries. It was good to use the old tools, which had honoured so many Ancestors, but the task was much easier with the knife, especially since Ibuso no longer possessed the strength she once had.

She paused for a second to suck the liquid from her fingers so that she could keep a firm grip on the skull. It would dishonour the Shaman's spirit and might anger it if she dropped the skull in the dust. Ibuso looked across to the men's hut and saw Kiguda sitting beside the doorway. It was Kiguda who had brought the knife back the village after a raid on a distant village to the North. Normally, the men would not have gone further than the adjacent villages a few miles away, but Kiguda had

Figure 10.7   Step 7: Is this only a story about aborigines?

University of Leicester. It doesn't take long to find he has excellent academic credentials, and the journey can be continued in the assurance that we are in secure, academic hands. In *Step 7*, I click on the yellow arrow next to '*You decide*'.

What on earth is this? This web page at:

http://www.virology.net/Tutorials/cow/cow3.html

...seems to comprise a short story about aborigines!

However, when I read the sentence '*A thin grey paste began to drip from the skull into the bowl. It had been easy to remove the Shaman's head from his neck using the broad steel knife*' it is almost impossible to avoid going on. Who said science was boring?

At the end of this fascinating story, in *Step 8*, I continue beyond this web page. I continue through *Steps 7–23* to find 16 web pages of well-presented, well-illustrated scientific information which shows the relevance of this Papua New Guinean story to our own life in the West, particularly in relation to BSE and CJD.

For example, at:

http://www.virology.net/Tutorials/cow/cow13.html

there is quantitative data about the changes in the rendering processes

Traditionally, MBM was prepared by a rendering process involving steam treatment and hydrocarbon extraction.

- a protein-rich fraction called 'greaves' containing about 1% fat from which MBM was produced
- a fat-rich fraction called 'tallow' which was put to a variety of industrial uses.

In the late 1970s, the price of tallow fell and the use of expensive hydrocarbons in the rendering process was discontinued, producing an MBM product containing about 14% fat in which the infectious material may not have been inactivated. As a result, a ban on the use of ruminant protein in cattle feed was introduced in July 1988:

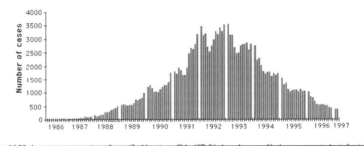

Figure 10.8 Step 8: Part of one of the web pages from Dr Cann's on-line tutorial

during the 1980s which almost certainly helped BSE to cross the species barrier to humans.

Although the story is now a little out of date by a couple of years, the conjunction of hard science, politics and cultural story-telling make this a pedagogic *tour de force*. Dr Cann has brought real creativity to an on-line tutorial, which entertains, enthuses and informs. For people who want to use the web for teaching, it is a *model of good practice*.

11
Example of a University Web Site

All pages in this example were visited between 1/10/99 and 27/7/01.

University web sites are usually well-produced. The web site described here for a university college at http://www.cant.ac.uk exemplifies some of the characteristics of a system which must cater for a medium-sized academic institution of about 10 000 students.

There is a *homepage* which, in this case has about eighteen major links, organised into four groups (Fig. 11.1).

- *Communications information* – how to find people, how to find the institution, how to get to it and find your way around it.

- *News and newly updated information* – here are the notices for one-off lectures and events such as the rag week and summer ball.

- *Basic functions of a university* – this section includes the mission statement of the University College and the links to the homepages of the faculties and departments. The link to the latter is a major route, followed frequently by both staff and students.

- *The frequently used links* are icons which represent direct links to the library, to a search engine for the University College web site, to a suite of general search engines, to the University College homepage and a help page.

Some of this information is conserved as a *frame* in other pages. For example, compare the two screens in Fig. 11.2. The binoculars (an icon meaning 'search') have been selected on the first screen, and take the reader to the search engine on the second screen. Although the second figure is a stage further on than the first, the *communications links* and the *frequently used links* have been retained. Incidentally, at the bottom of this web page,

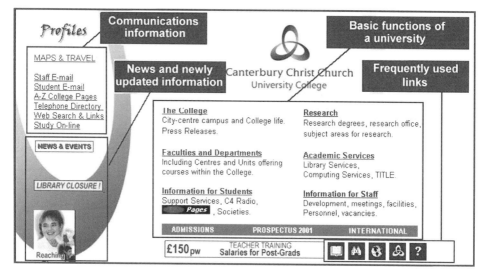

Figure 11.1 The University College homepage

we can see the email address of the person responsible for producing it. It is this person who deals with dead links and other anomalies.

From this page of *academic services*, if you click on the binoculars in 'frequently used links', you are taken to the search engine for the site. Such site-based search engines usually comprise a hidden index of all the web pages, where each word in the index is a hyperlink. When you insert your search word, the search engine matches it to a word in the index and then this hyperlink word takes you to the appropriate page.

Once the search engine has been invoked, the whole web site can be examined minutely. Some hosts only make information available through a *local search engine*. This can take up more time, and prevents the visitor from easily doing real, human browsing. You have to know what you are looking for, which doesn't allow for serendipity.

Although the layouts differ from one university to another, they all have these four groups of links. It is through the web pages that prospective students can get immediate information about availability of aspects such as *accommodation* and *courses* (Fig. 11.3).

Increasingly, tutors will use the internet for student support. For example, some excellent web pages for helping students with *study skills*, are freely available at:

http://www.cant.ac.uk/cware/list/list.htm and
http://www.cant.ac.uk/sssu/guides.htm

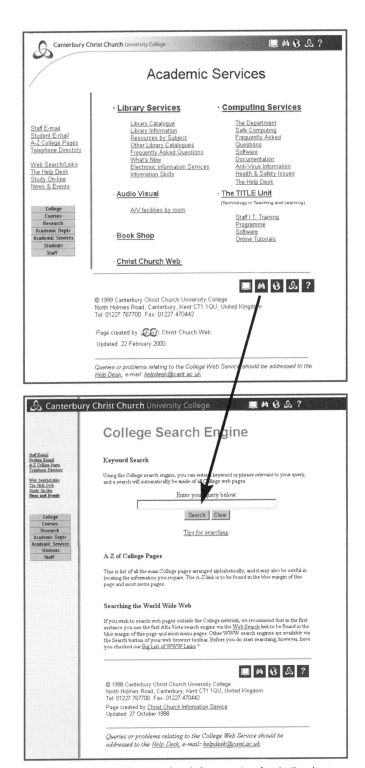

Figure 11.2   The next level down – Academic Services

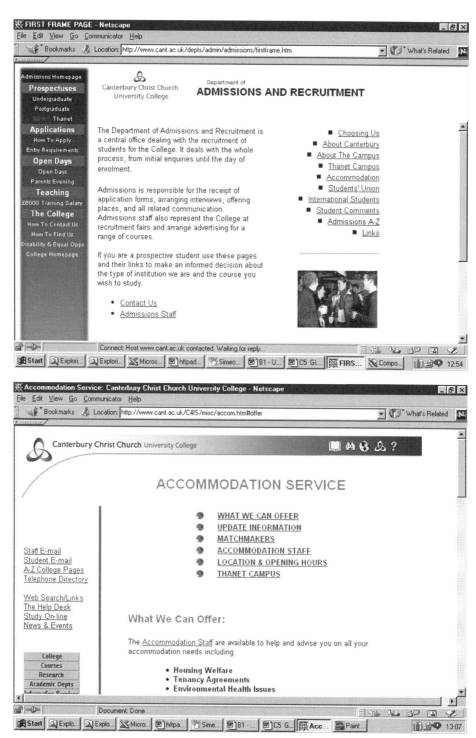

Figure 11.3 Typical information pages regarding availability of places on courses and availability of accommodation

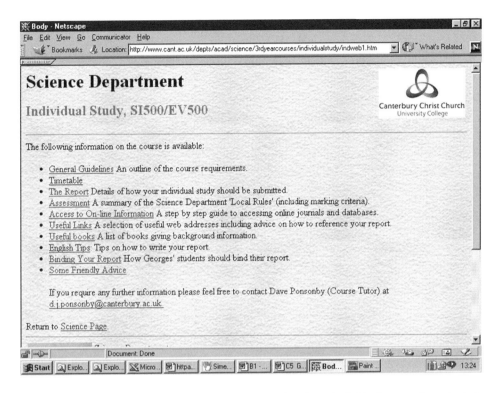

Figure 11.4 Web-based guidelines for students undertaking individual research projects in their final year

There are recurrent proposals in universities to make things available on the internet rather than giving students handouts. This saves money for the institution concerned but devolves the costs to the student, who will almost certainly have to print out copies of the web-based materials. To their credit, some tutors (who remember their impecunious student days) have refused to resort only to web-based materials and supply *both electronic and paper versions* of their materials. For example, at:

http://www.cant.ac.uk/depts/acad/science/3rdyearcourses/
individualstudy/indweb1.htm

shown in Fig. 11.4, as well as supplying paper handouts, tutors have used the web to give instructions on both carrying out, and then reporting on their final-year undergraduate research project. First-year students are given

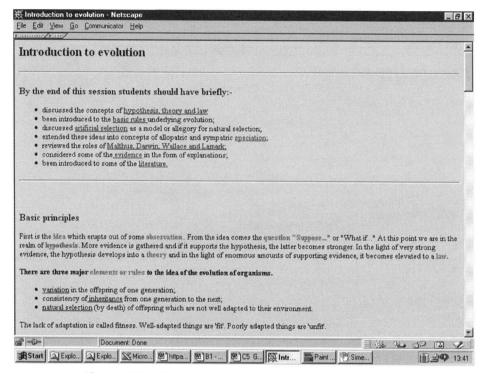

Figure 11.5 First-year course notes for a lecture on evolution

access to course notes which they can read on screen or print out
(Fig. 11.5) at:

http://www.cant.ac.uk/depts/acad/science/1styrcrses/environ/
georges/evol/Intrevol.htm

12

A Typical General Search – *the Wreck of the* Sea Empress

In the following example, all sites were visited several times between 5/11/00 and 27/7/01.

The *Sea Empress* was an oil tanker ship which recently grounded in Wales. During a world-class oil spill, enormous amounts of oil escaped onto a scenic and protected coastline. I would like to find out when the accident took place and some general information about the environmental impact. In *Step 1*, I go to the browser and find an appropriate search engine. In this case, I chose Excite:

http://home.netscape.com/uk/escapes/search/netsearch_1.html

For *Step 2*, the words '*Sea Empress*' are entered in the Search Box. This leads to a suite of web sites which relate to the subject at:

http://www.excite.co.uk/search.gw?lang = en&look = nscpuk%5Fuk&c = web.uknp&search = %22Sea+Empress%22&start = 10

From the composition of the URLs in the list, we can see that most of these web sites relate to work done by the University of Swansea. For example, in the middle of many of the URLs I see .swan.ac.uk where:

'uk' means 'United Kingdom'

'ac' means 'academic' (a university or institute)

'swan' means 'Swansea', which is the major city near to where the accident took place.

There are two clues that this is reliable information:

Figure 12.1   Step 1

Figure 12.2   Step 2

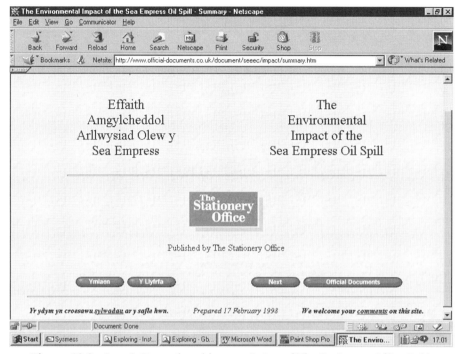

Figure 12.3 Step 3 (Reproduced by permission of The Stationery Office Ltd.)

1 It is produced by a recognised university.

2 There are many aspects to the same incident. This looks 'academic'. In fact, some of these sites look rather too academic for my original brief, so I shall search a little further afield.

In the list, there is a link to a web site entitled *Environmental Impact of the Sea Empress Oil Spill – Summary*. In the URL, I can see that it says *Official documents.co.uk*. Anyone can say that they have produced an official document, so care must be exercised. How do we know this information is good?

I decide to follow this line and when, in *Step 3*, I click on the link at:

> http://www.official-documents.co.uk/document/seeec/
> impact/summary.htm

I am taken to the welcome page of *The Stationery Office*. Further link-following seems justified.

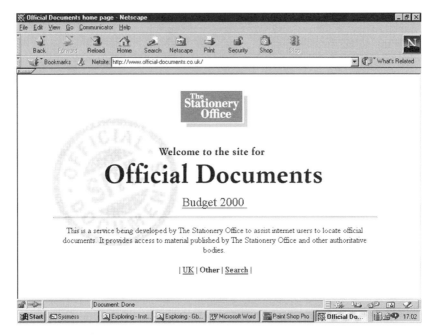

Figure 12.4   Step 4 (Reproduced by permission of The Stationery Office)

For *Step 4*, a click on the big button saying '*Official Documents*' might be useful. This looks promising. The search engine is evidently looking into the bowels of government information. However, what are we looking at? Is there information about the *Sea Empress* at the Budget 2000 web site? Ooops. When I come to:

> http://www.official-documents.co.uk/document/hmt/
> budget2000/hc346.htm

. . . I am gradually starting to understand that I have come to the budget proposed by the Chancellor of the Exchequer. This is not what was wanted. I have gone wrong somewhere and will have to go back a few steps. In *Step 5*, the back button of the browser on the top of the screen is used to go back through the pages I have looked at. Going back to Step 3, instead of clicking the hyperlink to 'Official Documents', I click on 'Next'.

By *Step 6*, real progress is being made. A map of the region appears at:

> http://www.official-documents.co.uk/document/seeec/
> impact/eng-map.htm

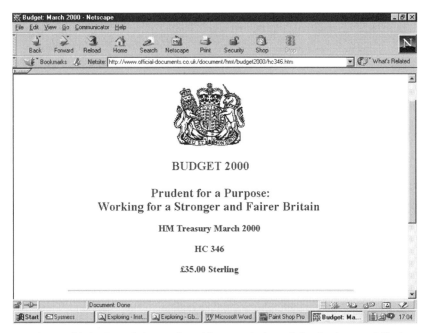

Figure 12.5 Step 5 (Reproduced by permission of the Stationery Office)

Unfortunately, a click on the title at the top of the page confusingly sends me back to the official documents page. However, scrolling down to the bottom of the page reveals another *Next* button which . . . in *Step* 7 gives me what I want at:

> http://www.official-documents.co.uk/document/seeec/
> impact/seeec-1.htm

The page notes that the *Sea Empress* foundered on *15 November 1996 in Milford Haven in south-west Wales*. This comprehensive document should tell me all that I need to know.

It is still useful to try to confirm whether this is *reliable* information. There are sometimes clues at the bottom of a web page but there is nothing significant here. However, on the first page is written:

> summary of the report on the spill prepared by the Sea Empress Environmental Evaluation Committee (SEEEC), an independent committee set up on 27 March 1996 by the UK Government.

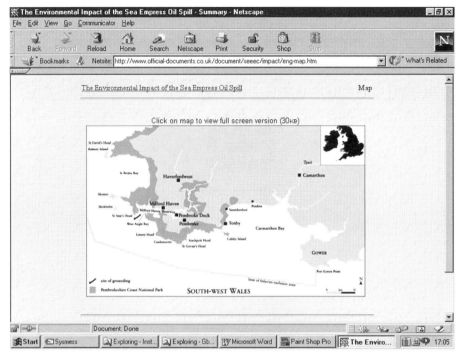

Figure 12.6 Step 6 (Reproduced by permission of The Stationery Office)

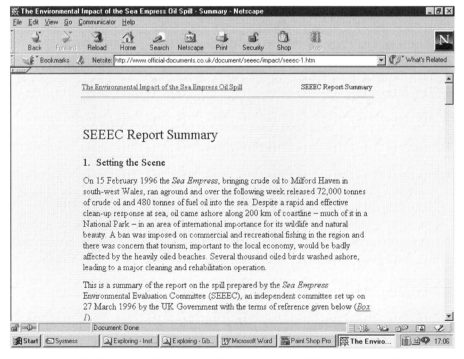

Figure 12.7 Step 7 (Reproduced by permission of The Stationery Office)

Again this seems to confirm that this is reliable information.

However, anyone can say they are a government committee. If it was *vitally* important, I would follow this further to verify that the Committee was a *bona fide* organ of government, possibly by writing by snail mail to the Department concerned.

13

Searching a Typical Government Site – mad cow disease

For this example, all sites were visited several times between 20/3/00 and 28/7/01.

Let us presume that basic information on *bovine spongiform encephalopathy* (BSE or 'mad cow disease') was required - in particular the incidence rates in Europe, and the relationship between consumer and polluter. BSE has been a problem in European agriculture and consumer affairs in Europe for many years. It has had grave repercussions, resulting in serious international disagreement and economic loss. As a European, I am particularly interested in the Western European context of this disease and decide that this should be my starting point.

In *Step 1*, the search words *European Union* are used in a search engine and lead to a page on health and consumer protection. The address of the site is:

> http://europa.eu.int/comm/dgs/health_consumer/
> index_en.htmI

It is possible to see from the final part of the address /index_en.htmI that this is the starting point, or index page of the whole site.

The hypertext link *Food Safety: from the Farm to the Fork* looks useful. A left click on this in *Step 2* jumps to a page at:

> http://europa.eu.int/comm/food/index_en.html

. . . which gives more details about the information which is available on this site.

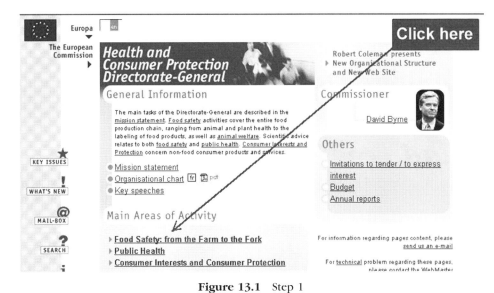

Figure 13.1 Step 1

Figure 13.2 Step 2

Many of the subjects look interesting but are not highlighted in blue and are not, therefore, hypertext-linked to anything. By contrast, in *Special Topics*, there is a hypertext link to BSE.

For *Step 3*, selecting this link takes me to a page at:

http://europa.eu.int/comm/food/fs/ah_pcad/ah_pcad_index_en.html

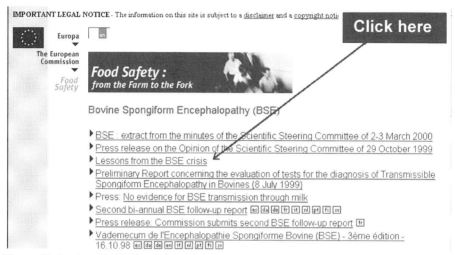

Figure 13.3   Step 3. NB The subject matter of this site substantially changed between 15/3/00 and 27/7/01. For example, on 4/4/01, the main issues had changed from BSE to *Foot and Mouth disease* and *Swine Fever*. However, on 4/4/01, the site search engine still revealed a huge range of documents relating to BSE, 5344 in total

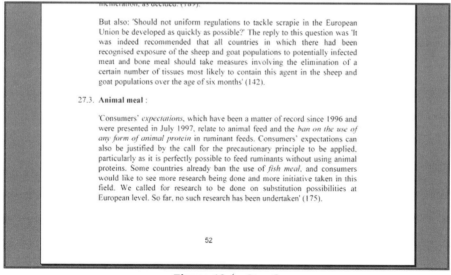

Figure 13.4   Step 5

... which focuses on BSE and yields a list of information including the *minutes* of committee meetings and press releases. The latter are often useful but the line which says *Lessons from the BSE crisis* looks particularly interesting. What might the lesson be?

TRENDS IN BSE
Confirmed cases of BSE (1) at 1/6/99

Country	Up to 1987	1988	1989	1990	1991	1992	1993	1994	1995	1996	1997	1998	1999 (2)	Total (2)	%
UK	442	2473	7166	14294	25202	37056	34829	24290	14473	8091	4334	3186	665	176501	
Isle of Man	-	6	6	22	67	109	111	55	33	11	9	4	-	433	
Jersey	-	1	4	8	15	23	35	22	10	12	5	6	-	141	
Guernsey	4	34	52	83	75	92	115	69	44	36	43	24	-	671	
Falkland	-	-	1*	-	-	-	-	-	-	-	-	-	1*	1*	
(UK)	446	2514	7229	14407	25359	37281	35090	24436	14560	8150	4391	3220	665	177747 (4)	99,43
B	-	-	-	-	-	-	-	-	-	-	1	6	1	8	-
DK	-	-	-	-	-	1*	-	-	-	-	-	-	-	1*	-
D	-	-	-	-	-	1*	-	3*	-	-	2*	-	-	6*	-
F	-	-	-	-	5	-	1	4	3	12	6	18	10	59	0,03
IRL	-	-	15**	14**	17**	18**	16	19**	16**	73	78	79	34	379	0,21
L	-	-	-	-	-	-	-	-	-	-	1*	-	-	1*	-
NL	-	-	-	-	-	-	-	-	-	-	2	2	2	6	-
P	-	-	-	1*	1*	1*	3*	14	14	29	30	106	63	260	0,15
EU excl. (UK)	-	-	15	15	23	21	20	40	33	114	120	211	110	722	0,40
CH	-	-	2*	2	8	15	29	64	68	45	38	14	16	299	0,17
Others (3)	-	-	-	-	-	-	1*	-	-	-	-	2	-	5	-
Total excl. (UK)	-	-	17	17	31	36	50	104	101	159	158	227	126	1026	0,57
World total	446	2514	7246	14424	25390	37316	35140	24540	14661	8309	4549	3447	791	178773	100%

* imported animals

** IRL.: figures including imported animals (5 in 1989, 1 in 1990, 2 in 1991 and 1992, 1 in 1994 and 1995)

(1) Statistical sources : European Commission, International Office of Epizootics (IOE) and Ministry of Agriculture, Fisheries and Food, UK (MAFF of April 1999)

(2) Provisional figures

(3) 1989 (Oman : 2) - 1993 (Canada : 1) - 1998 (Lichtenstein : 2)

(4) of wich 38975 borne after the banning of meat meal in bovine food in 1988 (sources MAAF)

Annual incidence : this criterion is more significant than the number of observed cases for the purpose of assessing the scale of the epizootic. It represents the quotient of the number of cases of BSE observed and the cattle population aged over 2 years in million head. The annual incidence was estimated in 1992 at 6653 the United Kingdom. In the other countries, it was estimated

	UK	P	IRL	F	CH
en 1996 :	1506,7	39,7	21,4	1,1	49,6
en 1997 :	816,2	41,1	22,4	0,5	41,9
en 1998 :	590,0	147,2	22,1	1,6	15,4

nvCJD : UK : 1995 (3) 1996 (10) 1997 (10) 1998 (16) 1999 (1) = 40 cases　　F : 1996 (1) = 1 case　　total : 41 cases

Figure 13.5　Step 6

This leads to a rather intimidating and technical page (not shown) which offers a choice of three links:

- Food safety and BSE

- Verbatim reports from meetings

- Participants.

The 'Verbatim Reports' might be worth looking at later; the 'Participants' are not really relevant to the purpose. For now, the crucial issue is *Food safety and BSE*. The next link offers a choice of three different languages and this link is selected as appropriate. *Step 4* takes me to an interim page which then suggests downloading a document called a .pdf file. The file extension .pdf tells us that the file should be looked at through a special viewer called Adobe Acrobat. This viewer is available free by an easy download from:

http://www.adobe.com/acrobat

Once you have the Acrobat reader and you call up a .pdf file, the reader prepares the file so that you can see it on the screen. Such .pdf files are relatively easy to *read on the monitor* and are economical with computer memory. When they are printed out, they appear as in the original document. It is possible to cut and paste small parts, or print out individual pages.

So, *Step 5* is to look at the document with the Acrobat reader.

A quick scan through the 58 pages of the document shows two pages of particular interest. One page contains text which looks exactly what was required. It relates the importance of the *precautionary principle for consumers*; this is a serious ethical consideration in a modern, consuming and polluting society.

In *Step 6*, I look at the other page. Here I find a table of incidence of *confirmed BSE cases* since 1989. This is exactly the kind of concrete information I was looking for. The search has been successful.

14

Using the Web for Bioscience Shopping

The internet makes *virtual shopping* easier, though sometimes financially risky. You can find and buy cut-price goods but need to make sure that you are not going to lose your money by fraud. Protecting against this usually involves common sense. For example:

- Keep your eyes and ears open as to which suppliers are *reputable*.

- Buy from a reputable source which is *not going to go into liquidation* immediately after you have paid.

- Pay by credit card, since your losses due to theft are recoverable from the credit card company. (However, keep in mind that credit card fraud is one of the most rapidly growing areas of crime!)

- Reliable suppliers are likely to have *encryption* systems in place to protect against credit card fraud.

Here is an example of a shopping expedition. A browser has already been used to find the *Field Museum* in Chicago in the USA at http://www.fieldmuseum.org 15/3/00–27/7/01. Assume we are interested in buying a model dinosaur skull.
The links to follow are:

to the main page	http://www.fieldmuseum.org/
to the store	http://www.fieldmuseum.org/store/default.htm
to dinosaurs in the store	http://www.fieldmuseum.org/store/store_dinosaur.htm
to details on a particular skull	http://www.fieldmuseum.org/store/dino_trrexkull.htm
to decide how I want to order	http://www.fieldmuseum.org/store/store_placeorder.htm
to actually place the order	http://www.fieldmuseum.org/store/store_orderform.htm

The sequence is shown in Fig. 14.1. Only parts of each page are shown.

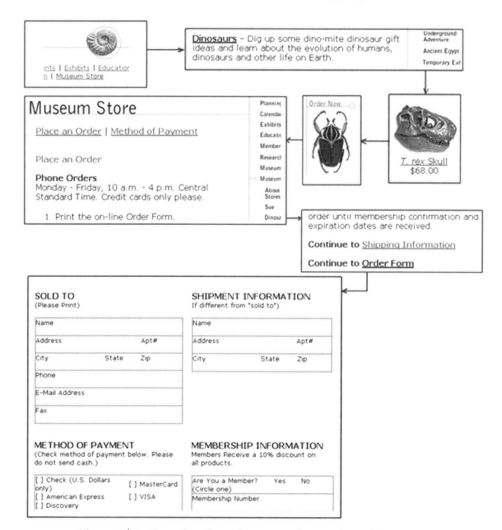

Figure 14.1   Procedure for making a purchase at the Field Museum

When shopping on the internet, observe the following rules:

- shop only from firms you think you can *trust*
- investigate the *refund* and return policies before you order
- check the small print – *buyer beware*!
- make sure the server is *secure*, usually indicated by a padlock on the page that requires your credit card details

- contact the company *by email* if there is anything about which you are unsure

- watch out for *hidden extras* such as postage and packing; customs and tax charges can be very high

- keep a *record* of the transaction, the email address and the web site.

Wentk (1998) covers many *consumer aspects* of the internet.

The Field Museum shopping trip has features which are common to most internet shopping. There is a pressure on the sellers to make things as easy as possible. Despite this, it is still possible to come across examples of poor web design which don't address the needs of the user carefully enough.

15
Whaling –
Freedom of Enquiry

All web sites quoted in this example were visited several times between 20/3/00 and 28/7/01.

The internet can be a medium for carrying debate for government, non-government organisations and citizens. In this example, the net is used to consider some of the arguments for and against whaling. Let us presume I have an assignment in which the question is *'Should whaling be banned completely or should it be treated as a sustainable economic resource?'* My general knowledge tells me that certain countries such as Norway and Japan want whaling to be allowed, perhaps under a regime of resource management; I also know that NGOs such as Greenpeace have taken positive action against whaling ships by, for example, placing small boats between the whales and the whaling ships.

I hope to find solid evidence of

- *positions adopted* by the players in this discussion
- *some facts* which are being used in the arguments.

15.1 THE INTERNATIONAL NETWORK FOR WHALING RESEARCH

The search begins with *Step 1* where the browser is used to access a search engine, in this case the now defunct AltaVista at:

http://www.altavista.com

The search word *Whaling* is inserted in the search box.

AltaVista gives 200 sites relating to this subject.

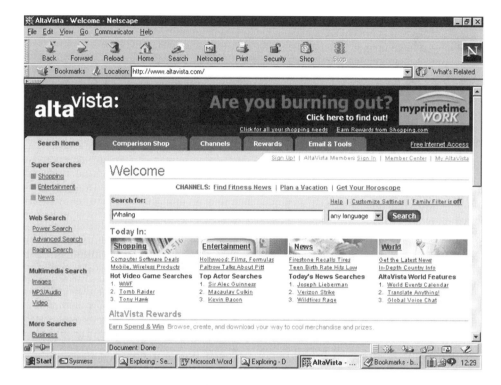

Figure 15.1   Step 1

Most of the sites on the first page concern whaling museums, which were not the objective. However, at the bottom of the first page is a site for the *International Network for Whaling Research*, so in *Step 2* I click on this link and it takes me to the Introduction section of the web site at:

http://www.ualberta.ca/~inwr/INWR.html#Introduction

Here, it says that the Network was formed at a meeting in Denmark in 1990. It appears to be an academic web site, since it includes fisheries scientists, economists and political scientists. Its role is to spread information about whaling research amongst interested parties. It also intends to promote research. I now have two questions about this site:

1 How active is it now?

2 What is its position with respect to whaling? Is it completely independent?

> **9. International Network for *Whaling* Research**
> Welcome to the International Network for **Whaling** Research. Click on Section Title to Access. Introduction. Introduction to the Network. Members ...
> URL: www.ualberta.ca/~inwr/INWR.html
> Translate ▣ Related pages

Figure 15.2 Step 2

The index show some recent papers from 1999, so it appears to be still 'alive'. An email to the Editor confirms this.

On the evidence so far it certainly seems to be an independent, academically based network. The newsletter might give some clues, so this link is followed by clicking on 'News'. At the top of the browser screen, I see:

http://www.ualberta.ca/~inwr/INWR.html#News

The #News tells me that this information is on *this* web page (as signified by the #). Although I have found the information through a hypertext link, I can also see it by merely scrolling down the main page.

The impression that this is an academic site is confirmed when I look at the last few lines of the web page and find out that the INWR Secretariat is based jointly at a Canadian university, the Whaling Research Project in Germany and a university in the Netherlands.

In *Step 3*, I investigate the site further.

In the news section is an eye-catching headline which states:

March 6, 1998: World Whaling Communities Unite to Assert Their Rights

Most of the information one sees in the press is about pressure on whaling communities not to go whaling, so this looks useful. I click on the hypertext-linked headline and then see a site which is familiar to regular users of the internet (Fig. 15.4):

http://www.ualberta.ca/~inwr/INWR-Digest.htm

The search engine has been told to follow a link to a particular web page, but this page no longer exists so there is a dead link. The search engine therefore sends a message back to say that the web site cannot be found on that server, namely the server at the University of Alberta.

In this case, we are fortunate, as those responsible for the site have been careful to give an alternative URL.

In *Step 4*, this alternative URL takes us to a list of documents,

http://www.ualberta.ca/~inwr

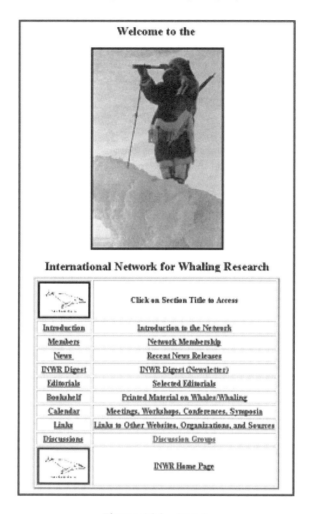

Figure 15.3 Step 3

...but there is no information, and I don't know which one to choose. Since the date for the headline shown above was March 6, for *Step 5* I choose an article dated as close to this as possible, i.e. *Selected INWR Digest Editorials 18-Feb-1998*.

http://www.ualberta.ca/~inwr/INWR-Editorials.html

In the most recent editorial, for June 1997, is the statement:

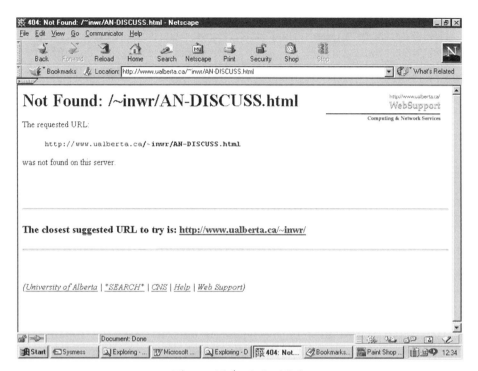

Figure 15.4 A dead link

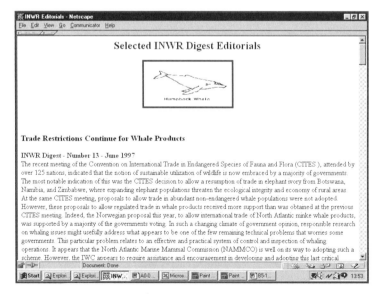

Figure 15.5 Step 5

However, these proposals to allow regulated trade in whale products received more support than was obtained at the previous CITES meeting. Indeed, the Norwegian proposal this year, to allow international trade of North Atlantic minke whale products, was supported by a majority of the governments voting. In such a changing climate of government opinion, responsible research on whaling issues might usefully address what appears to be one of the few remaining technical problems that worries some governments.

and I have the vague feeling that a stance is being taken.

The editorials seem to be in chronological order, so it might be worth taking a look at the first, to see if any clues as to stance can be found there. The passage quoted below from Digest 6 August 1998 seems to indicate that this site has not always been completely neutral and should therefore be handled with caution (note my emphases in the quotation).

Thus, in the context of IWC discussion, aboriginal/subsistence whaling is tolerated (because it is believed, unlike industrial whaling, to be technologically simple and not to involve monetized exchange) whilst all non-aboriginal/commercial whaling, for the opposite reasons, cannot be tolerated. Much of this exaggerated fear *centres on the Marxian belief that allowing the exchange of goods for money necessarily results in evil outcomes – a strange belief for anti-whaling groups whose annual revenues often total in the millions of dollars.* For many of these organizations, representing inter alia, animal rights, animal welfare, and assorted environmental causes, the annual IWC meetings provide an excellent arena for pronouncing their ideas before a large assembly of journalists and government officials. It is ironic that as the whaling industry has shrunk in monetary terms to a fraction of its former size, so the whale protection industry has grown (both in terms of money and personnel) to many times the size of the remaining whaling operations. The problem, as many social scientists know only too well, is that the 'worth' of a human activity cannot be assessed in monetary (or strictly economic) terms alone, and failure to understand the multiple values that sustainable whaling represents to small coastal communities comprises one of the more serious misunderstandings of the protectionists. The problem with the easy accumulation of wealth is that it has the potential to encourage greed, and a consequent reluctance to share resources. *The whalers see no problem with sharing the whales (a renewable resource after*

all) with those who have other purposes in mind. Unfortunately, the
protectionists, appearing to manifest all the intolerance of recent
converts to a new ideology, lack any such generosity of spirit.

My feelings about 'stance' are confirmed when I go back to the main page
and investigate INWR Digest, where the four latest issues all have editorials
which adopt a pro-whaling standpoint.

This site shows us that scientists do not stand outside society. *Every*
scientist is a citizen and will have a view which will colour their
perspectives. Aside from expecting scientists to do their work in an
appropriately methodological way, we cannot expect them to behave as if
they are outside of, or beyond, humanity. The crucial thing is that we, the
readers, need to recognise the limitations of the ways in which scientists
might present their views to us.

15.2   THE NORWEGIAN GOVERNMENT POSITION

Perhaps it would be useful to take this investigation further.

From general knowledge, I know that Norway has a historic interest in
whaling. In *Step 1*, the search engine called Google is used to follow up the
search word *Norway*:

http://www.google.com

Google gives me a long list of possible sites and I see one which refers to
consular information. This looks as though it might be useful, since it may
give me solid, government-based information. However, when I follow up
this link, it becomes evident that this is directed at US citizens visiting
Norway, which is not at all what I was after.

In *Step 2*, I scan through the rest of the links and come across one on the
bottom of page 2 which offers information on the *state of the environment*
in Norway. This looks promising.

When I arrive at this site in *Step 3*, it seems to contain lots of useful
environmental information despite being in Norwegian. Since this might be
a useful site for the future, I bookmark this web page before I lose it.

http://www.grida.no/soeno95

However, I must not be distracted from the job in hand. I now need to
verify the quality of the information. Looking down at the left-hand side of

Figure 15.6 Step 1

Figure 15.7 Step 2

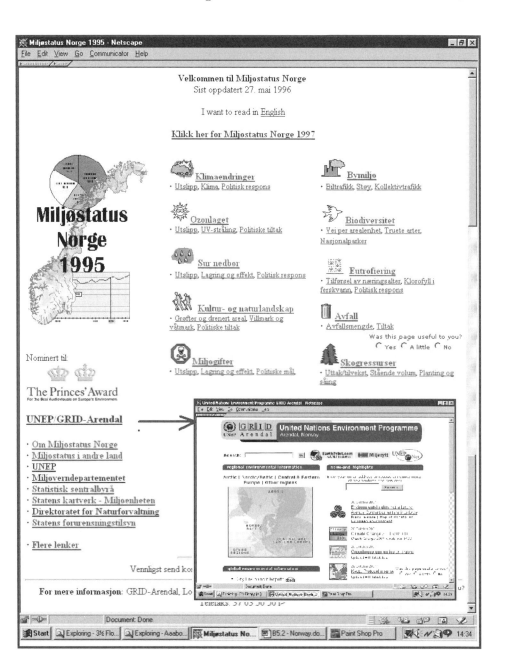

Figure 15.8 Step 3

the page, I can see links, for example to the *Norwegian Pollution Control Authority*, which give me reassurance. There is also an 'about' link. These are always useful, since they often tell the visitor about the aims and context of a web site.

I decide to visit this quickly just to ensure that this site is governmental. The link 'about SOE Norway' takes me to a *United Nations Environmental Protection Page* based at Arendal in Norway:

http://www.grida.no/index.htm

Overcome by curiosity at the wealth of information which seems to be available, I bookmark it and then decide to look at the links to *Maps and graphics – complete library* at:

http://www.grida.no/index.htm

I am now faced with a list of links to various environmental issues and regions. As a European, I click on *Europe* and am then given 16 pages of maps showing everything from regional geography to sulphur dioxide deposition on Scandinavia:

http://www.grida.no/db/maps/prod/level1/70902.htm

It may not be about whales, but it looks like extremely useful information for other environmental essays I may need to write, and I didn't get too distracted. As I accumulate these serendipitous sites, I realise that I need to keep my Bookmark (or Favorites) file in good order so that I can search through it easily.

Back to the job in hand. Bearing in mind that whales are not fish, in *Step 4* I click on *fish resources* in the bottom right-hand corner and am given detailed information about the Norwegian fisheries – for example in the 1930s, there were about 120 000 fishermen and now there are less than 30 000:

http://www.grida.no/soeno95/fishery/fishery.htm

Following up the links here does not take me to anything concerned with whales, so I go back to the state-of-the-environment page and in *Step 5* try the Ministry of Fisheries, where there is immediately a link to something to do with whales at:

http://odin.dep.no/fid/engelsk/index-b-n-a.html

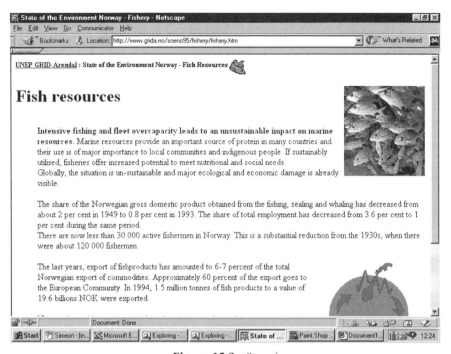

Figure 15.9 Step 4

Figure 15.10 Step 5

Step 6 takes me to a brochure giving facts about whales in Norwegian waters which contains some excellent biological information at:

http://odin.dep.no/fid/engelsk/p10001872/p10001953/008001-120003/
index-dok000-b-n-a.html

I also learn that:

Facts about Whales in Norwegian waters

Balance in the food chain

Whales, of which there are approximately 75 species, are a part of the marine ecosystem. They either eat fish, or they compete with fish for their food. Most whales need to eat the equivalent of approximately five

Minke whale
Bottlenose whale
Narwhale
Killer whale
Pilot whale
Beluga
Porpoises and dolphins
Bottlenose dolphin
Common dolphin
White-beaked dolphin

Figure 15.11   Step 6

Whale meat has been a part of the daily diet for over 1,000 years. It tastes delicious and is healthy. Recent research indicates that the oil in whalemeat and blubber substances has a preventive effect on cardiovascular diseases, among others.

Following the link to *Whaling, Step* 7 reveals a map of minke whale distributions and the page notes that:

Norwegian Minke whales are hunted using ordinary small boats, about 60 feet long on average, all of which are licensed for whaling. These boats are re-rigged for the whaling season and are equipped with modern explosive harpoons. Meat and blubber are refrigerated on board and are brought to shore for processing and packing. Whaling takes place in the Norwegian zone of the North Sea and along the entire coast of North Norway, eastwards to the Barents Sea and off Spitsbergen. In addition, some whales are caught off Jan Mayen.

However, I still don't have a view of Norwegian government attitude to whaling, so I skip back to the Ministry of Fisheries page and follow up two links – one of a report to the Norwegian Parliament and the other being the opening address of the Fisheries Minister to a Conference. Neither of these do more than mention whaling:

The proposals for the amended Act relating to the right to participate in fishing, whaling or sealing suggest that a licence to trade can be

Figure 15.12   Step 7: Density distribution map of minke whales in the North Atlantic

granted if the applicant has carried out commercial fishing, whaling or sealing on or with a Norwegian vessel for at least three of the past five years.

I suspect that the official Norwegian position is mentioned in earlier reports which have not yet been put on the web. The *Norwegian trail has dried up*, and I would need a strong incentive to take it further through the normal channels of academic study such as inter-library loans and computerised database searching. This could be disappointing but the bonus is that my Bookmark (Favorites) file now contains some excellent ancillary web sites.

15.3   THE INTERNATIONAL WHALING COMMISSION

Although I am beginning to get a 'feel' for what is going on, I haven't really managed to obtain much solid factual information. There have been some

Figure 15.13   Step 1

references to the *International Whaling Commission (IWC)* and I have heard about them on the news, so perhaps they have a web site with data? I am short of time so I need to be certain that I can get there in as few hits as possible. For *Step 1*, I decide to use a meta-search-engine called Webcrawler and use 'International Whaling Commission' as my search words.

This works well and *Step 2* presents a useful set of links which come from a range of sources. Some of these sites seems to relate to political aspects; perhaps I should have begun with a more specific search.

Academic work is often *iterative* – in other words you have to return to the same issues repeatedly and modify what you have already done. Although political aspects were found in the earlier searches, it might be advisable to return to some of these sites later and, therefore, I Bookmark (or Favoritise) the site. However, *I have to keep my eye on the ball . . .* and at this moment I want data.

There are a number of interesting sites here. One relates to whale population dynamic modelling. Clicking on this link at:

http://www.stat.washington.edu/raftery/Research/
Whales/whales_papers.html

. . . takes me to a list of high-powered articles such as '*A Bayesian Framework and Importance Sampling Methods for Synthesizing Multiple Sources of Evidence and Uncertainty Linked by a Complex Mechanistic Model. Geof H. Givens Ph.D. Dissertation, Department of Statistics, University of Washington, Seattle, 1993*'. Overwhelmed by the titles alone, I give up on this possibility. There is a link to an IWC conference in 1996 but I would like something more recent and fundamental.

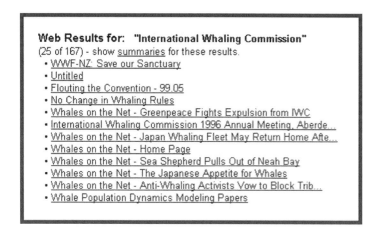

Figure 15.14   Step 2

Figure 15.15   Step 3

Scrolling down the list, I look for something extremely specific or extremely general, and come to a hyperlink which says 'links to other sites'. Note that at the top of the list, I am told that this page represents items 76–100 out of a possible 167.

Step 4 comprises selecting a link which leads to the world-famous *Scott Polar Research Institute* (*SPRI*) in Cambridge:

http://www.spri.cam.ac.uk/links.htm

Surely they must have a link to the IWC? Yes – they do! However, the SPRI one seems to be a useful site which carries some of the cachet of Cambridge

Scott Polar Research Institute

Links from SPRI to other Websites

---

1. Some other useful websites in Cambridge

- Cambridge University Home Page
- British Antarctic Survey is linked from our home page.
- Cambridge Arctic Shelf Programme UPDATED
- International Whaling Commission, Histon, Cambridge
- Cambridge Online City - community information UPDATED
- Be a tourist in Cambridge ! - webpages from Cambridge City Council UPDATED

Figure 15.16 Step 4 (© Scott Polar Research Institute, University of Cambridge)

University. Academic information here is probably reliable, so I bookmark this site and in *Step 5*, press on to the IWC homepage at:

http://ourworld.compuserve.com/homepages/iwcoffice/iwc.htm

The design of the IWC homepage leaves something to be desired. Nevertheless, it gives me some options. There is a hyperlink on *conservation and management* at:

http://ourworld.compuserve.com/homepages/iwcoffice/
iwc.htm#Conservation

. . . which sends me to a page of useful prose which contains further links. There are some problems with the design and performance of this site but I persevere.

One phrase which catches my eye says 'Click here for current population estimates'. I want to follow this up because the whole argument over whaling ultimately rests on the two issues of:

- *sustainability*

- *ethics of hunting*

and even the latter depends on the former. If the numbers are so small that survival is threatened because of problems arising from issues such as

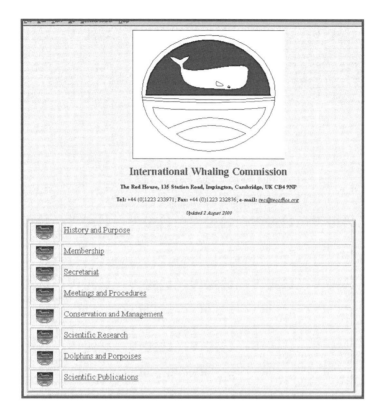

Figure 16.17 Step 5

genetic bottlenecking, then all the other aspects become relatively insignificant. In genetic bottlenecking a population becomes so small that there is too little genetic diversity. Populations need genetic diversity in order to survive future changes in the environment. Only when the populations are sufficiently large can we justify harvesting or hunting.

When I see the table on current population estimates in *Step 6*, there seems to be good news and bad news at:

http://ourworld.compuserve.com/homepages/iwcoffice/Estimate.htm

The *good news* is that there appear to be reasonable breeding populations of most whales.

The *bad news* is that the blue whale appears to be on the verge of extinction.

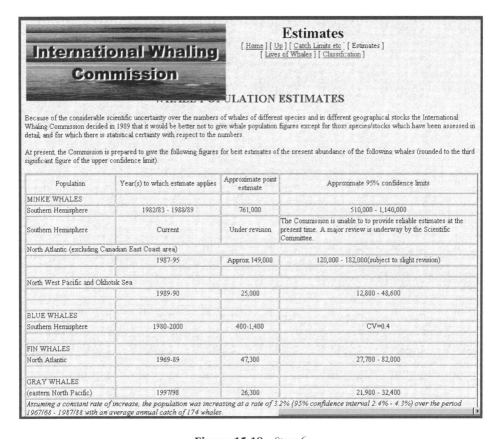

Figure 15.18   Step 6

A possible population density of *400 blue whales* looks low – and the coefficient of variation (a measure of how variable the results might be) looks high. So there could be a lot fewer or a lot more.

In *Step 7*, I find information on catch limits for subsistence whaling which could be central to the final argument of my assignment.

What is the IWC doing about controlling whaling and keeping the process sustainable?

When I scroll down below the table found in Step 6, I discover a section concerning a 'Comprehensive Assessment' (or moratorium) which came into being in 1982. When I take *Step 8* by clicking on the hyperlink *'agreed to a pause in commercial whaling'* at:

http://ourworld.compuserve.com/homepages/iwcoffice/
Schedule.htm#PARA10e

Catch limits for aboriginal subsistence whaling

The Commission reviewed catch limits of stocks subject to aboriginal subsistence whaling. The following limits have been agreed:

Bering-Chukchi-Beaufort Seas stock of bowhead whales (taken by Alaskan Eskimos and native peoples of Chukotka) - The total number of landed whales for the years 1998, 1999, 2000, 2001 and 2002 shall not exceed 280 whales, with no more than 67 whales struck in any year (up to 15 unused strikes may be carried over each year).

Eastern North Pacific gray whales (taken by those whose "traditional, aboriginal and subsistence needs have been recognised") - A total catch of 620 whales is allowed for the years 1998, 1999, 2000, 2001 and 2002 with a maximum of 140 in any one year.

West Greenland fin whales (taken by Greenlanders) - An annual catch of 19 whales is allowed for the years 1998, 1999, 2000, 2001 and 2002.

West Greenland minke whales (taken by Greenlanders) - The annual number of whales struck for the years 1998, 1999, 2000, 2001 and 2002, shall not exceed 175 (up to 15 unused strikes may be carried over each year).

East Greenland minke whales (taken by Greenlanders) - An annual catch of 12 whales is allowed for the years 1998, 1999, 2000, 2001 and 2002 (up to 3 unused strikes may be carried over each year).

Humpback whales taken by St Vincent and The Grenadines - for the seasons 2000 to 2002, the annual catch shall not exceed two whales

The Scientific Committee is continuing its investigation of potential new management regimes for aboriginal subsistence whaling.

Figure 15.19 Step 7

International Whaling Commission

International Convention for the Regulation of Whaling, 1946
SCHEDULE
As amended by the Commission at the 51st Annual Meeting 1999, and replacing that dated September 1998

PLEASE NOTE: THIS IS NOT A FACSIMILE EDITION. IT HAS BEEN MODIFIED STYLISTICALLY FOR EASE OF USE ON THE WEB. THE OFFICIAL SCHEDULE IS THE PRINTED VERSION AVAILABLE FROM THE OFFICE OF THE COMMISSION (Click HERE for details).

Figure 15.20 Step 8

...it takes me to the Schedule of the main international agreement on whaling. At *Step 9*, I am now where I originally wanted to be, in the meaty parts of the subject.

In *Step 10*, I look through the footnotes relevant to para.10(e) at:

http://ourworld.compuserve.com/homepages/iwcoffice/
Schedule.htm#FOOTPARA6)

I notice that in relation to taking whales...

The Governments of Japan, Norway, Peru and the Union of Soviet Socialist Republics lodged objections to paragraph 10(e) within the prescribed period. For all other Contracting Governments this paragraph came into force on 3 February 1983. Peru withdrew its objection on 22 July 1983.

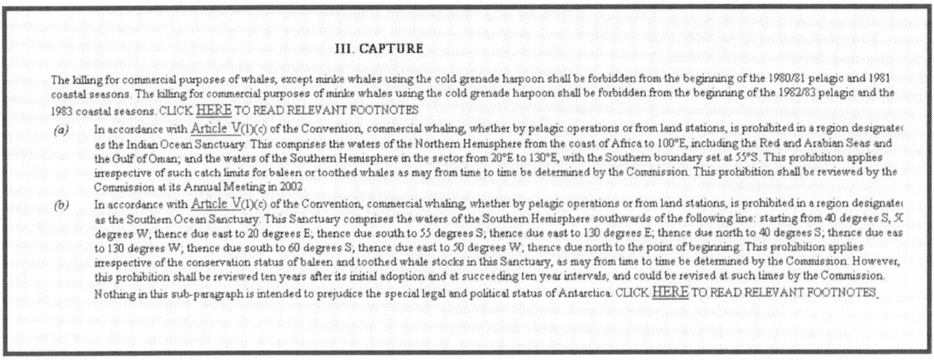

III. CAPTURE

The killing for commercial purposes of whales, except minke whales using the cold grenade harpoon shall be forbidden from the beginning of the 1980/81 pelagic and 1981 coastal seasons. The killing for commercial purposes of minke whales using the cold grenade harpoon shall be forbidden from the beginning of the 1982/83 pelagic and the 1983 coastal seasons. CLICK HERE TO READ RELEVANT FOOTNOTES

(a)   In accordance with Article V(1)(c) of the Convention, commercial whaling, whether by pelagic operations or from land stations, is prohibited in a region designated as the Indian Ocean Sanctuary. This comprises the waters of the Northern Hemisphere from the coast of Africa to 100°E, including the Red and Arabian Seas and the Gulf of Oman; and the waters of the Southern Hemisphere in the sector from 20°E to 130°E, with the Southern boundary set at 55°S. This prohibition applies irrespective of such catch limits for baleen or toothed whales as may from time to time be determined by the Commission. This prohibition shall be reviewed by the Commission at its Annual Meeting in 2002.

(b)   In accordance with Article V(1)(c) of the Convention, commercial whaling, whether by pelagic operations or from land stations, is prohibited in a region designated as the Southern Ocean Sanctuary. This Sanctuary comprises the waters of the Southern Hemisphere southwards of the following line: starting from 40 degrees S, 50 degrees W, thence due east to 20 degrees E; thence due south to 55 degrees S; thence due east to 130 degrees E; thence due north to 40 degrees S; thence due east to 130 degrees W; thence due south to 60 degrees S, thence due east to 50 degrees W; thence due north to the point of beginning. This prohibition applies irrespective of the conservation status of baleen and toothed whale stocks in this Sanctuary, as may from time to time be determined by the Commission. However, this prohibition shall be reviewed ten years after its initial adoption and at succeeding ten year intervals, and could be revised at such times by the Commission. Nothing in this sub-paragraph is intended to prejudice the special legal and political status of Antarctica. CLICK HERE TO READ RELEVANT FOOTNOTES.

Figure 15.21   Step 9

The Government of Japan withdrew its objections with effect from 1 May 1987 with respect to commercial pelagic whaling; from 1 October 1987 with respect to commercial coastal whaling for Minke and Bryde's whales; and from 1 April 1988 with respect to commercial coastal sperm whaling.

The objections of Norway and the Russian Federation not having been withdrawn, the paragraph is not binding upon these Governments.

This partially answers one of my previous questions about Norway's role in whale hunting. For both the Russians and the Norwegians, it is evidently still a seriously live issue.

Conclusion

This has been a long journey but in relation to whaling, it has allowed me to identify:

- a group of people who have *a particular view* with which I may or may not agree,
- governments which have *not agreed* to follow the same protocols as others,
- the *official web site* of one of these governments,
- sites which can give me lots of *extra information*, including data, about whaling,
- the main *international agreement* relating to whale conservation management.

When doing an assignment or an exam answer, you should constantly refer back to the question which was posed, to make sure that you have stayed on target. The original question was:

Should whaling be banned completely or should it be treated as a sustainable economic resource?

To answer this question I need to know:

1 the *nature of the resource*, i.e. what it is composed of and how big it is,

2 the *nature of those who might exploit it*, i.e. who they are, whether they can be controlled or can control themselves and at what rate the resource will be used.

The university library can supply the major parts of 1, though I have encountered some web sites here which will give me ancillary information. The web sites will make a major contribution to item 2. I will probably use both these sources of information to make my own back-of-an-envelope calculations. From an environmental perspective, I know from my training as an ecologist that the arguments are in favour of strongly controlled hunting. Most of the ethical scientific arguments hinge on *the way* in which certain animals are hunted.

Finally, the question implicitly asks for my *opinion* and I, the writer, might want to bring my own moral or ethical position into the final parts of the answer. I should certainly have enough information to come to my own conclusions and make a sensible argument.

16

Searching Via Research Groups

All web sites quoted in this example were visited several times between 10/10/00 and 28/7/01.

Many research groups offer information on the web concerning their work. Often, before the esoteric research details, there are some pages of introductory background which can be useful for a novice biologist. In the more advanced parts of such web pages, it is *wise to be wary*, since in a few sites, researchers might be using the web to 'talk-up' parts of their work that are not fully completed or have not been verified. Remember that everyone is trying to make a career for themselves.

If the web page makes reference to *peer-reviewed* research published in appropriate journals then the information is almost certainly reliable. The example shown below uses the example of a good, scientifically reliable research group to show how the pages of a research pages can be investigated.

Suppose:

1 it is necessary to find information on some aspect of *cellular biology relevant to medical treatment*,

2 a previous assignment comprised an essay on hormone pollution, and in doing the latter, I became interested in *hormonal control in general*.

From my general knowledge, I suspect that one of the cancer research foundations might have some relevant information. Consequently, as *Step 1*, I use a search engine such as Excite to find the homepage of the UK Imperial Cancer Institute at:

http://www.icnet.uk

Figure 16.1 Step 1

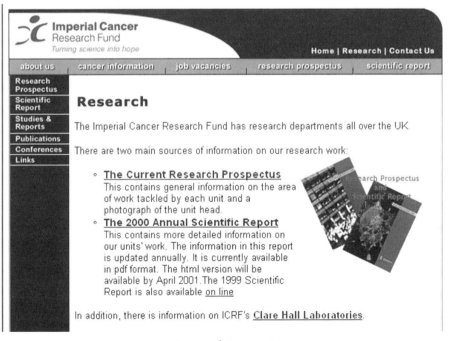

Figure 16.2 Step 2

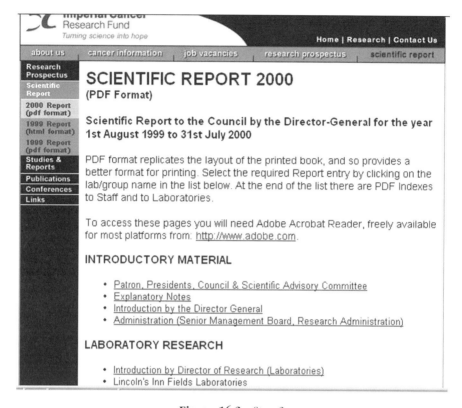

Figure 16.3 Step 3

In *Step 2*, I click on *Research*:

> http://www.icnet.uk/research/index.html

... and in *Step 3*, a click on *2000 Scientific Report* takes me to:

> http://www.icnet.uk/research/scirep2000/pdf/index.html

A quick scan down the page for the keyword *endocrinology* shows a title *'Molecular Endocrinology Laboratory'*. This looks useful and in *Step 4*, leads to a .pdf file which describes the background to the work of this particular research group:

> http://www.icnet.uk/research/scirep2000/pdf/index.html

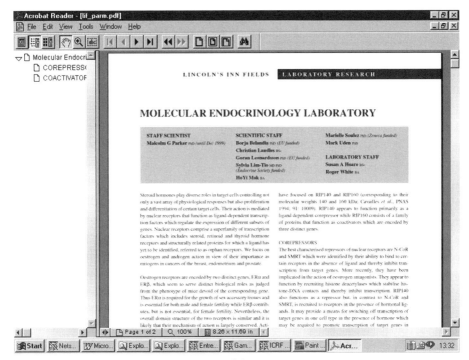

Figure 16.4 Step 4

Many of the specialised bioscience research groups have such summary pages which introduce their work.

In this case, I find out that there are two genes which control *oestrogen reception*. One of these genes controls the growth of accessory sexual tissue while the other controls female fertility. Steroid hormonal action appears to be closely linked to other signalling pathways. Deacetylases may fix histone proteins onto the DNA so that transcription is stopped. *RIP40* is a repressor which in (say) target 1 will switch off transcription in the presence of a hormone, where the latter might promote transcription in (say) target 2.

Reading on, it is evident that this site, like many other research group sites, yields state-of-the-art information directly relevant to the initial questions.

Many research labs produce on-line *synopses* of their recent and current work and this can be a fruitful hunting ground in which to search for up-to-the-minute information to include in an assignment. The bad news is that it will take a little initiative in the hunt, and some time, before you can find exactly the most relevant material. The good news is that making the effort will earn many more accolades than merely going directly to a source such as the *Encyclopaedia Britannica.* However, remember what was said earlier about *being wary.* Most novice biologists would not want to go so deeply into the subject. More experienced students should *read around the subject sufficiently to be constructively critical* of the results that are being presented. Keep in mind that such information is not peer-reviewed. As in the examples covered in Chapter 15, a research scientist or team might have their own reasons for presenting their results in a certain light.

Part 3
Final destinations

So where does all this leave us? All that is needed is some appropriate targets to get things moving...

17

An Eclectic List of Web Sites

Although the most obvious way to search is to use an appropriate search engine, sites still have to be critically evaluated by the student visitor and that takes time, knowledge and skill.

In this chapter, I have tried to compile a list of useful web sites which seemed to offer reliable information relevant to a bioscientist. There is a comment on each web site, and a note of the date when it was first visited. All these sites were last visited and still viable on 31/7/01. Contrary to expectations, the mortality of these sites is low. Of 540 appropriate web sites selected for inclusion in Chapter 17 in July 2000, only 5% had ceased to exist twelve months later. Some of these apparent losses were merely due to servers being down at the time of the search – the following day, the links were working again. Some losses related to revisions at a single university site and one encyclopaedia site. Judicious URL dissection showed that most of these sites were still alive elsewhere on their servers.

Of the 200 sites similarly included in Chapter 18, only 4.5% had disappeared twelve months later, and for similar reasons to those described above. No failed sites have been retained in the following lists.

Sites were selected on the following criteria:

- academic worth

- solid, factual content

- free access without payment

- probable stability

- reasonable download time.

Such a list of sites is quite *personal*. Each professional biologist would probably come up with a different list. Nevertheless, the following lists demonstrate web sites which are relatively *reliable* and *stable* and which offer *solid information*. By consulting these sites, a student could save a lot of searching time. Within the bounds of my opinion, no questionable, doubtful or scientifically sites have been included, except for a few which were intentionally listed to show the dangers, and which are flagged accordingly. I have not given a star or other rating to the sites. First, it was difficult to differentiate the sites since all those listed were good. Secondly, it is not possible to predict the appropriateness of a site for a particular user.

In Chapter 18 is a list which is based on the AQA Advanced-level syllabus and curriculum for biology taken by school students aged 16–18 in the English educational system. These are equivalent to *Gymnasium* students on the continent of Europe. In this second list, I have indicated the age range of the students who might use it. This age recommendation can only be approximate, because, for example, very able students might be able to interrogate the material at a high level. In general, I have tried to avoid duplication of web sites on the two lists. It is expected that secondary school teachers would supervise their pupils' use of these lists.

For both lists, each web site was visited and inspected. Many web sites were visited which could not be included in either of these lists for a variety of reasons.

It is disappointing that so few universities apply the ethos of *open source* to their curriculum materials. Consider a student in Scotland who uses web-based curriculum materials posted by the Hypothetical University of Amble. It is unlikely that the university would have made any money from this student if they had made charges for access, and it is unlikely to lose any money by virtue of the student having free access. With respect to curriculum materials, British universities still seem to operate in the monetaristic Dark Ages of the 1980s, where isolationism and competition are the watchwords, and cooperation and synergy obstructed.

When transcribing a URL to start a search, do not include the text which is in italics.

STARTING POINTS

http://www.britannica.com   *If you are not sure of your subject, then this is where you begin. You will get background and links here, but for undergraduates, this is not enough. You must go further 30/3/01*

http://bubl.ac.uk   *Information service for the UK academic community, offering Dewey-indexed links to many sites and hosted at Strathclyde*

University. An essential starting point for an undergraduate assignment. For example, interested in neurophysiology, I was led to interactive models of a dissection of the brain of a fly at http:// flybrain.uni-freiburg.de/Flybrain/html/contrib/1999/antpaper/vrml2 antlobe/index.html 7/3/01

http://www.sheffcol.ac.uk/links/Science/Biology *Classified, annotated and searchable database of 20000 educational web sites of which 500 are scientific. Gained 2000 BECTA/Guardian award for post-16 resources. Excellent set of links for all aspects of biology – with comments 16/1/01*

http://www.biozone.co.uk./links.html *This is an excellent source of a wide range of biological links. If you are stuck, search here 18/3/01*

http://biology.about.com/science/biology *This makes a useful alternative to Biozone or Accessexcellence, with a US bias 16/1/01*

http://www.ultranet.com/~jkimball/BiologyPages/T/TOC.html *Biology pages from the senior author of a US biology textbook 18/3/01*

http://www.accessexcellence.org/AB/GG *A wide range of biologically relevant topics can be pursued via this site presented by the National Health Museum in the USA 18/3/01*

http://esg-www.mit.edu:8001/esgbio/chapters.html *A free on-line biology textbook hosted by MIT, one of the world's most prestigious universities 18/3/01*

http://gened.emc.maricopa.edu/bio/bio181/BIOBK/BioBookTOC.html *A free, well-linked, on-line biology book 18/3/01*

http://biocrs.biomed.brown.edu/Books/TOC-Biology-TLS.html *On-line support, well illustrated, related to a text published by Prentice Hall. There are lots of teaching resources here 18/3/01*

http://biome.ac.uk/about *This site, hosted at the University of Nottingham comprises links to a wide range of academically respectable resources; for example, it linked to 12 sites which could provide solid data on foot and mouth disease such as that provided by the Office International des Epizooties (OIE) at http:// www.oie.int/eng/en_index.htm 10/3/01–1/12/01*

http://hjs.geol.uib.no/Introduction/content.html-ssi *This unusual and interesting site, based at the University of Bergen in Norway, offers a wide range of geological and biosciences information based on the interests of the web authors. For example, there is a research report of a comprehensive collaborative project on the condition of lakes in the eastern Mediterranean at http:// hjs.geol.uib.no/Professional/Projects/Eastern_Mediterranean.html-ssi 13/11/00–1/12/01 (Fig. 17.1).*

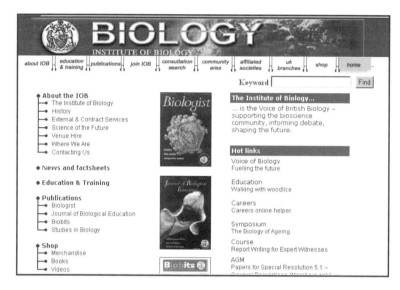

Figure 17.1   Website of the Institute of Biology, www.iob.org.uk (© Institute of Biology)

http://www.bbc.co.uk/education/asguru/biology   *In consultation with the chief examiner of AS Level, this BBC site focuses on difficult aspects of the curriculum in biology. However, downloads seem to take a long time 18/3/01*

http://phylogeny.arizona.edu/tree/home.pages/search.html   *The Tree of Life hosted by the University of Arizona. This is a good starting point if you are concerned with relationships – for example, taxonomy or evolution 30/3/01*

http://www.boxmind.com   *On-line educational consultancy which offers links to many other sites. Keywords Evolutionary Biology led to 5 sites, together with a brief description of what the site comprised. Guest lectures are offered but were not working at the time of my visit 7/3/01*

http://sdb.bio.purdue.edu/Other/VL_DB.html   *Virtual library for the Society of Developmental Biology, maintained by Purdue University in the USA 7/3/01*

http://www.clearinghouse.net/cgi-bin/chadmin/viewcat/Science___Mathe matics/biology?kywd++   *This is a potentially useful source site hosted by the University of Washington, with lots of links to useful information for students 7/3/01*

http://www.clearinghouse.net/index.html   *Yahoo-style searching centre with many academic links 7/03/01*

http://www.ase.org.uk/bilink.htmlhttp://www.ase.org.uk/bilink.html
Important list of biological resources from the Association for Science Education in the UK 9/11/00

http://discovery.com *Cluttered site but gives access to lots of resources. Teaching plans lead to a range of biolinks 9/11/00*

http://www.madsci.org/libs/libs.html *Useful starting point based at the Washington University Medical School. Called 'Madscience', it offers a wide range of science links and resources 23/01/01*

http://www.fathom.com/index.jhtml *Collaborative international project by universities and museums to give access to academic information. You have to register, which is a little inconvenient, but it offers lots of links and is worth the effort 11/3/01*

http://www.bbc.co.uk/science/humanbody/articles/general/brainstory_perception.shtml *BBC science site with information and links; for example, this page shows curiosities of human perception 2/02/01*

http://www.bbc.co.uk/education/asguru/biology/03organssystems/index.shtml *This is another BBC site. There are masses of useful information here, especially for school pupils, but the downloads take a long time 18/3/01*

http://www.thinkquest.org/about/index.html *Thinkquest is a non-profit association which offers a home for lesson plans and student work. Users need to note that they may be looking at the work of students – for example the brave effort at http://library.thinkquest.org/21939/harmful_substances_pollute_the_a.htm 18/3/01*

http://www.glencoe.com/sec/science/biology/bls/index.html *A US publisher of biology texts offers basic information and links related to the units in the books 18/3/01*

http://genbiol.cbs.umn.edu/PIF_files/PIF_Docs/PIF_R_Files/library/gbweb.html *This is a set of biology links developed by the University of Minnesota which do not give the generalised information available in textbooks. So if you want something advanced or unusual, you might start here 5/3/00*

http://vm.cfsan.fda.gov/~frf/biologic.html *This is a site offered by the Food and Drug Authority in the USA which lists many other sites of biological relevance. However, many of the links are to aspects such as journals and societies to which one must subscribe 1/4/01*

http://www.sciam.com/explorations *Scientific American articles on the internet. Well worth browsing if you have an essay to do 30/3/01*

http://www.the-scientist.com/homepage.htm *Homepage of the US journal 'The Scientist'. Via a free subscription procedure, you get access to lots of readable articles. Well worth browsing if you have an essay to do 30/3/01*

AGRICULTURE

http://www.defra.gov.uk   *Homepage of the Department for Environment, Food and Rural Affairs in the UK. This is a good source of information but most people in the business think it is a doubtful agglomeration of interests when it comes to scientific objectivity. Handle carefully 18/3/01*

http://www.ecotrop.org   *There is comprehensive information at this site but be wary as the author has a particular point to make about biotechnology in agriculture 5/3/00*

www.agriculture.gouv.fr/accueilv4f.htm   *Solid information on French agriculture, from the French Ministry – in French 3/2/01*

http://www.nfu.org.uk/info/f&ml.asp   *Foot and mouth – the farmers' perspective from the web site of the National Farmers Union in the UK 1/4/01*

http://www.ext.vt.edu/resources   *The Virginia Cooperative shares information, particularly concerning scientific aspects of agriculture 1/4/01*

ALGAE

http://www.umich.edu/~phytolab/GreatLakesDiatomHomePage   *Comprehensive set of detailed images of diatoms of the Great Lakes of North America 13/11/00*

http://hjs.geol.uib.no/diatoms/Biology/index.html-ssi   *University of Bergen site concerning biology, taxonomy and morphology of diatoms 13/11/00*

http://hjs.geol.uib.no/diatoms/Introduction/index.html-ssi   *Introductory but nevertheless detailed and reliable University of Bergen information on diatoms 13/11/00*

ANATOMY

http://www.microscopy-uk.org.uk/about.html   *Microscopy-UK is a not-for-profit site which offers a wide range of images, mostly microscopical. Although the images are copyrighted, the site could well be a useful source of information for an essay or practical report 1/11/00* http://www.exploratorium.edu/bodies/index.html *The Exploratorium offers medical images of the human body, including magnetic resonance images 11/3/01*

http://biology.miningco.com/science/biology/msub21.htm *Index page for virtual dissections 18/3/01*

http://www.danke.com/Orthodoc/flexhandfing.htm *Human anatomy in relation to pain reporting, as used by remedial massage therapy 22/3/01*

http://www.le.ac.uk/pathology/teach/va2/index.html *Virtual autopsy offered by the University of Leicester in the UK. There are a number of case studies. Excellent if you have the stomach for it 1/4/01*

http://www.science.lander.edu/rsfox/invertanat.html *Library of dissection guides to accompany an undergraduate course at Lander University 1/4/01*

http://www.science.lander.edu/rsfox/limulus.html *Anatomy of Limulus, a living arthropod fossil 3/2/01*

AQUACULTURE

http://AquaTIC.unizar.es *Electronic magazine for aquaculture (in Spanish) 8/11/00*

AQUATIC BIOLOGY

http://www.latene.com *Ireland-based source of maps and books (commercial) on fisheries and gas etc. from a consortium of companies. No obvious freebies 12/1/01*

http://www.utas.edu.au/docs/plant_science/HAB2000/abstracts/docs/James_Kevin_J.html *Conference abstract with subject of toxic blooms 20/11/00*

http://biology.rwc.uc.edu/HomePage/BWS/planktonkey/phytozoo.html *Useful pictorial plankton key 22/01/01*

http://perso.wanadoo.fr/erb *French site concerning rivers and lakes. There is lots of technical information here (in French) 3/6/00, 11/3/01*

http://boto.ocean.washington.edu/eos/hydro_data.html *Aquatic models (as images) of the Amazon, hosted by Washington University 18/7/01*

http://daac.gsfc.nasa.gov/CAMPAIGN_DOCS/OCDST/poster_supplement.html *Seawifs is a global international marine monitoring project 5/3/00* http://www.eaufrance.tm.fr   *Water in France, in English. There is lots of good information here 1/4/01*

http://www.rivernet.org   *European rivers network – scientific colla-boration with worldwide interests. The rivers information tends to be in the local language 1/4/01*

http://www.iawq.org.uk   *The International Water Association based in the UK offers a wide range of information on water management 1/4/01*

BEHAVIOUR – SEE NEUROPHYSIOLOGY

BIOCHEMISTRY

http://pdb.weizmann.ac.il/scop/data/scop.1.html   *Structural classifica-tion of proteins. There are some wonderful images on this university site 18/3/01*

http://biocrs.biomed.brown.edu/Books/Chapters/Ch%208/DH-Paper.html   *Copy of the original paper by Watson and Crick on DNA structure 18/3/01*

http://www.biology.arizona.edu/biochemistry/problem_sets/energy_enzymes_catalysis/energy_enzymes_catalysis.html   *Enzyme biochem-istry on-line objective question test hosted by the University of Arizona 18/3/01*

http://q10.com.au/page1.htm   *Australian page about the membrane-soluble cofactor Q10 1/4/01*

http://faculty.washington.edu/ely/coenzq10.html   *US page about the membrane-soluble cofactor Q10 1/4/01*

http://invader.bgsm.wfu.edu/overview.html   *Biochemical research into flavoproteins at Wake Forest University. The introductory parts contain some good, basic information 22/7/01*

http://www.vuw.ac.nz/~teespitt/essentials/index.htm   *Illustrations from a general textbook on biological chemistry. There are some useful visual materials here 23/7/01*

http://www.worthpublishers.com/lehninger   *Valuable companion web site for Lehninger's Principles of Biochemistry. Students have open access. Tutors need to register in order to be given a password which gives access to the tutor-based materials 1/7/01*

http://www.the-scientist.com/yr2001/jul/research_010723.html   *Recent research on prions described in an article in the Scientist magazine 27/7/01*

BIODIVERSITY

http://www.biodiversity.org/Handbooks_eng.htm *Biodiversity Conservation Information System. This is a poorly presented site with lots of important information including free on-line publications such as an extensive conservation procedures manual. In my opinion, the web designer needs counselling 17/3/01*

http://www.nhm.ac.uk/science/projects/worldmap/index.html *Interactive world map of biodiversity, hosted by the Natural History Museum of London 10/3/01*

http://www.nrmc.demon.co.uk/bdpro *Good, clear diversity and community statistics package as beta download 23/01/01*

http://viceroy.eeb.uconn.edu/EstimateS *Species richness and lots more, including Shannon and Fisher & Simpson's indices. Not working 23/01/01*

http://homepages.together.net/~gentsmin/ecosim.htm *Simulation for null model analysis in community ecology. Compares hypothetically modelled communities with real ones. Free download (look below the picture!) 23/01/01*

http://nhsbig.inhs.uiuc.edu/www/chi.html *Ecology modelling programs. Looked useful but the Predation program would not unzip properly. Is there a problem with the Chinese language here? 23/01/01*

http://detritus.inhs.uiuc.edu/general_stats/alpha.lst *Useful ecological programs and information including radio-telemetry software and a list of Fisher's alpha for different numbers of species and individuals 23/01/01*

http://www.biology.ualberta.ca/jbrzusto/krebswin.html *Programs and resources available based on Ecological Methodology by C. J. Krebs 23/01/01*

http://www.kovcomp.co.uk/downl2.html *Ordination, clustering and diversity programs 23/01/01*

http://www.ecocam.com/software/ *Checklist of Software for field biologists 23/01/01*

http://www.landcare.cri.nz/science/biodiversity *Biodiversity issues presented by New Zealand research institute. For example, there are pdf information sheets on poisonous plants in NZ as well as the usual museum lists of holdings 22/01/01*

http://www.biodiversity.org.uk/*ibs* *Biodiversity on the internet. Source site which leads to others, including images and plotting data. This site links to a biodiversity search engine at http://ibs.uel.ac.uk/ibs/ globalsearcher/ 28/02/01* http://www.biodiversity.nl   *Netherlands site with lots of information on the subject, and choice of languages. There are lots of links 7/3/01*

BIOGRAPHY

http://dir.yahoo.com/Science/Biology/Biologists   *The Yahoo site which has links to biographies of about 20 famous biologists 5/3/00*

http://www.ucmp.berkeley.edu/history/lamarck.html   *Californian university tells you about Lamarck, the evolutionist 5/3/00*

http://www.ucmp.berkeley.edu/history/wegener.html   *...and about the man who proposed continental drift 5/3/00*

http://kidscience.about.com/kids/kidscience/msub8.htm?once = true&   *This site might be for kids but it gives useful and linked biographical information on a host of scientists 5/3/00*

BIOLOGICAL PROBLEMS

http://www.bsereview.org.uk   *Food Standards Agency Review of BSE 11/11/00*

BIOMES AND HABITATS

http://www.dfait-maeci.gc.ca/english/foreignp/arctic.htm   *The Arctic Council is a forum of eight countries with interests in the Arctic 17/3/01*

http://www.envir.ee/baltics/frame1.htm   *The Baltic Sea – an official site 1/4/01*

http://www.livinglakes.org/bodensee   *The Bodensee in Germany is described in English 1/4/01*

BIOTECHNOLOGY

http://www.accessexcellence.com/AB/GG/#Anchor-Biological-3800   *Big list of links to biotech web sites 2/3/01*

http://www.whiterose-bio.com/html/biotechnology_web_at_the_universities.html   *British university consortium list of links 26/3/01*

http://www.frtr.gov/matrix2/sitemap.html   *US Federal Remediation Technology web site on innovative technologies for hazardous waste remediation 1/4/01*

http://www.nhm.ac.uk/zoology/ciliate/index.htm   *Protistans in sewage works 1/4/01*

http://www.sewagetreatment.cc   *US-based information exchange site which documents sewage works installation disasters 1/4/01*

BOTANY IN GENERAL

http://www.york.ac.uk/res/ecoflora/cfm/ecofl/index.cfm   *Ecological flora of the British Isles with information on a suite of over 130 ecological and morphological characteristics, relating to vice-county distribution in Britain, European distribution by country, mycorrhizal associations and fungal diseases. This is a valuable database at the University of York 25/7/01*

http://sunflower.bio.indiana.edu/~rhangart/plantmotion/plantsinmotion. html   *Movies of plants doing things; needs the Quicktime plug-in 9/11/00*

http://miavx1.muohio.edu/~journalcwis/compass97/kiss.html   *Personal essay on plants in space 9/11/00*

http://.taggart.glg.msu.edu/bot335/335syl.htm   *Lectures of a university botanist with some excellent pages including 'Origin of Life' 9/11/00*

http://.taggart.glg.msu.edu/bot335/protista.htm   *Introduction to the Euglenophyta, Chrysophyta, Pyrrophyta. University site 9/11/00*

http://jxb.oupjournals.org/cgi/content/abstract/50/339/1541   *Abstracts of papers in the Journal of Experimental Botany 5/3/00*

http://fcbs.org   *Florida Council of Bromeliad Societies. Everything you ever wanted to know about bromeliads ('parlor plants') 30/3/01*

http://biology.rwc.uc.edu/HomePage/BWS/planktonkey/phytozoo.html   *Useful pictorial plankton key offered by Raymond Waters College in the USA 22/01/01*

http://www.nhm.ac.uk/botany/clayton   *Homepage of the Clayton Herbarium in the London Natural History Museum 1/4/01*

http://www.rrz.uni-hamburg.de/biologie/b_online/e00/contents.htm   *On-line botany textbook hosted by the University of Hamburg. Material on this web site was originally presented in German but is gradually being presented in English. This looks like a good general resource 27/7/01*

http://www.botany.utoronto.ca/ResearchLabs/McCourtLab/Pages/ABA.html   *University researcher's homepage on hormones 27/7/01*

CELL BIOLOGY

http://www.ucmp.berkeley.edu/alllife/eukaryotalh.html   *Although   pre-sented in a palaeontological context, there is lots of free, basic information on cell biology at this University of California site 18/3/01*

http://www.cellsalive.com/index.htm   *Interesting site on cell biology, with masses of accessible, visual and stimulating information. The provenance is not clear but the information looks good 5/3/00*

http://vlib.org/Science/Cell_Biology/cell_cycle.shtml   *On-line   virtual library of cell biology 26/3/01*

CHARITIES AND TRUSTS

http://www.wellcome.ac.uk   *This is the web site of a major biomedical research charity which produces, for example, an excellent series of publications called 'Topics in International Health' directed particu-larly at developing countries. It's a shame they couldn't have been made available, on-line, for free 19/07/01*

CHEMISTRY FOR BIOLOGISTS

http://www.chem.arizona.edu/massspec   *Clear course notes introducing mass spectrometry from the University of Arizona 2/5/99*

http://www.umass.edu/microbio/chime/index.html   *Web site for Chime, which is a free program to show molecular structure in three dimensions 2/5/99*

http://www.chem.leeds.ac.uk/Project/MIME.html   *This site concerns the use of the internet for a variety of chemistry-based activities such as molecular modelling. There are valuable links to a host of highly regarded institutions and packages. The acronym of the project is MIME (Multipurpose Internet Mail Extension) 2/5/99*

http://www.mdli.co.uk/cgi/dynamic/newsmagazine.html?uid = $uid&key = $key   *Web site for MDL, a life sciences and chemistry informatics company based in California since 1978. There is a magazine which contains some general articles. The site offers a free download of ISIS Draw, a molecular modelling package 2/5/99*

http://www.knowledgebydesign.com/tlmc/tlmc.html   *Useful site with a wide range of chemistry-based learning materials including computer graphics, on-line exercises to increase chemical common sense, a software library, an interactive Periodic Table game and a discussion of safety issues etc. 2/5/99*

http://www.wellesley.edu/Chemistry/chem.html *Chemistry software and tutorials, including statistics available freely from this university in the USA 2/5/99*

CLIMATE CHANGE See Fig. 17.2

http://www.ipcc.ch *Intergovernmental panel on climate change 22/2/01*
http://www.unfccc.de *Text version of UN framework convention on climate change 22/02/01*

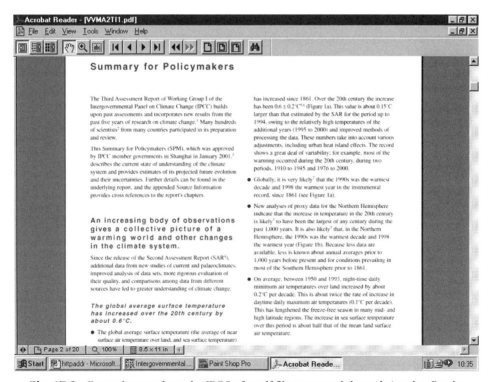

Fig. 17.2 Example page from the IPCC of a pdf file presented through Acrobat Reader

COMPUTING AND SOFTWARE

http://www.icann.org/general/abouticann.htm *International   Corporation for Assigned Names and Numbers 20/11/00*

http://www.symantec.com/avcenter/vinfodb.html *Information on computer viruses including a virus encyclopaedia 22/01/01*

http://www.pnas.org/help/search_help.dtl *Clear statements of searching rules for the Proceedings of the National Academy of Science. These rules apply to most searches on most search engines 25/02/01, 3/4/01*

http://www.internetsurvey.co.uk *Data from an ongoing survey of net use (c. 4342 respondents). Females outnumber males by 58 to 42% 7/3/01*

http://vil.mcafee.com/hoax.asp *Mcafee is a renowned virus protection company. Here they supply free information on viruses and virus hoaxes 7/3/01*

http://www.Vmyths.com *Virus hoaxes are described here. There is an alphabetical list. Half an hour spent on this site could save you hours later in dealing with hoax virus messages 7/3/01*

http://www.symantec.com/avcenter/venc/data/family.pictures.html *Another antiviral and software supply site. This one tells you about a virus hoax called 'Family Pictures' 8/3/01*

www.safekids.com *Healthy site which presents children's web-safety rules in an accessible way. It is worth a visit by adults too 7/03/01*

http://www.ispa.org.uk *Internet Service Providers Association. Find out about ISP activities here 7/3/01*

CONSERVATION

http://www.unep-wcmc.org *Homepage of the World Conservation Monitoring Centre 10/3/01*

http://www.iucn.org *Homepage of the International Union for the Conservation of Nature (World Conservation Union) 25/02/01*

http://iucn.org/themes/ssc/siteindx.htm *Site index for the Species Survival Commission. This page points to many sources of conservation information, including for specialist groups 25/02/01*

http://www.iucn.org/themes/ssc/redlists/rlindex.htm *Red Data List of threatened animal species (Fig. 17.3) 22/02/01*

http://www.iucn.org/themes/ssc/guidelines.htm *Draft guidelines for inclusion on the Red Data Lists. There is a lot of excellent, solid information in this paper 25/02/01*

Figure 17.3   The IUCN site at http://www.redlist.org which lists the most threatened species

http://www.unep-wcmc.org *UNEP threatened plants searchable database can be found here, based on the 1997 Red Data List of endangered plants 22/02/01*

http://www.silvanus.co.uk *Woodland trust which has undertaken woodland conservation for 14 years. Mainly practical advice 22/01/01*

http://www.wastewatch.org.uk *Recycling and materials conservation web site. The organisation is part-funded by the government and offers solid and reliable information on waste disposal including links and details of 1000 recycled products available for purchase in the UK 22/01/01*

http://www.recycle.mcmail.com *Official government advice on recycling. Lots of links and useful information here 22/01/01*

http://www.kentwildlife.org.uk *Local conservation survey 1/12/00*

http://nrs.ucop.edu/reserves/carpinteria.html *University of California Salt Marsh Reserve at Carpinteriam, California. Local but detailed information is available at this web site 22/01/01*

http://nt1.ids.ac.uk/eldis/eldsea.htm *Search engine page for the British Library for Development Studies. This is an excellent starting point for good information relating to conservation and sustainability. It gives summaries of the activities of the organisations to which it points 25/02/01*

http://www.schutzstation-wattenmeer.de   *Wadensee German site concerning environmental protection in Schleswig-Holstein 11/3/01*

http://www.nasco.org.uk/html/about_nasco.html   *North Atlantic Salmon Conservation Organization is an international organisation 11/3/01*

http://zebu.uoregon.edu/1998/phys162.html#top   *Excellent suite of course notes on energy conservation from the University of Oregon 18/3/01*

http://www.snf.se/english.cfm   *Swedish Society for Nature Conservation – in English. Lots of information here 1/4/01*

http://www.rspb.org.uk   *Royal Society for the Protection of Birds (UK) offers interesting links and excellent photographs 22/01/01*

http://www.nt-education.org/nteducation/index.html   *The UK National Trust education web site. Unfortunately it all requires money 22/01/01*

http://www.bdt.org.br   *Brazilian site offering global information on sustainable environment 5/3/00*

http://www.tatihou.com   *National park in north-west France – in French 1/4/01*

http://129.12.13.7/iibmircen/interreg_project.htm   *Scientific studies in relation to land reclamation from the Channel Tunnel 2/2/01*

CURIOSITIES

http://www.nhm.ac.uk/interactive/science-casebooks/bm   *Science casebooks at the Natural History Museum UK which include an account of the scientific hunt for the Beast of Bodmin Moor 1/11/00*

DEVELOPMENTAL BIOLOGY

http://www.luc.edu/depts/biology/dev.htm   *This developmental biology site links to many other useful sources of information 23/3/01*

http://www.exploratorium.edu/exhibits/embryo/embryo.html   *Shows similarity in embryos of fish, chickens, lizards, humans 1/11/00*

http://www.visembryo.com/baby/week40.html   *Development of the human embryo is shown at this Visible Embryo site. There is a relatively cheap subscription for anything more than the most general information 30/3/01*

DIVERSITY – SEE BIODIVERSITY

EARTH SCIENCES FOR BIOSCIENCES

http://www.british-geological-survey.co.uk   *British Geological Survey as a shopping site. It appears that everything has to be paid for, though there is a table of asbestos resources in the world, e.g. a table showing bauxite production would cost £10 – iniquitous 12.1.01*

http://gened.emc.maricopa.edu/geology/earth_sci.html   *Wide range of links from the Estrella Mountain Community College, USA 26/3/01*

http://www.geog.ouc.bc.ca/physgeog/contents/table.html   *Okanagan College in Canada offers useful information, including a glossary, on aspects of physical geography, including biogeography and evolution 1/4/01*

http://arnica.csustan.edu/common/geologic_time.htm   *Geological time scale chart 1/4/01*

ECOLOGY

http://www.colostate.edu/Depts/Entomology/courses/en570/readings.html   *Chemical ecology paper on apple maggot chemical attractants 10/3/01*

http://www.envirolink.org/mkzdk/texts/gaia.html   *Introduction to the Gaia hypothesis (actually a book review) on a visually stunning, privately maintained site 11/3/01*

http://benthic.sc.edu/benthic/BIOL/Courses/BIOL301/Wethey   *Valuable course notes on quantitative ecology from the University of South Carolina 18/3/01*

http://jan.ucc.nau.edu/~doetqp-p/courses/env470/Lectures/lec38/Lec38.htm   *Lecture notes from North Arizona University with useful images 18/3/01*

EDUCATION IN BIOLOGY

http://www.bbc.co.uk/education/asguru/biology/intro.shtml   *This is the BBC learning web site which offers the core information on biology for pupils aged 11-15. It is a good revision site for these pupils 1/4/01*

http://redtail.eou.edu   *Eastern School of Education and Business – a site search revealed 85 lesson plans for school 24/10/00* http://www.worldbank.org/html/edi/toronto/post/ra-eetc/sld005.htm *(Slides – long download but useful information about an environmental education project failure) 11/11/01*

http://www.nceet.snre.umich.edu   *Environmental education resources on the internet. Good site for starting off 7/02/01*

http://www.naturenet.com   *Wisconsin-based environmental education resources 11/3/01*

http://www.accessexcellence.org   *Pooled resources on the web, from school and college teachers 18/3/01*

http://www.fed.cuhk.edu.hk/~johnson/teaching/teaching.htm   *Good discussion and resources site maintained by the University of Hong Kong 18/3/01*

http://www.nhptv.org/natureworks   *A range of educational resources are offered by Squam Lakes Natural Resources Center, USA 26/3/01*

http://library.thinkquest.org/20465/laws.html   *An example of award-winning work produced by students in the USA 26/3/01*

http://www.visionlearning.com/about   *Visionlearning is a free site where teachers can pool resources. It looks aesthetically pleasing 26/3/01*

http://www.learn.co.uk/guardianarchive/edulearn/education_resources/lesson_packs/science/default.htm   *This is a free educational resource web site offered by the Guardian newspaper in the UK. Although a bit underdeveloped at the time of writing, it promises to be an excellent resource 11/3/01*

http://trace.ntu.ac.uk/kotn/teachers.htm#science   *Links to science resources for younger children, particularly noting UK sites 3/4/01*

http://www.educate.org.uk/teacher_zone/classroom/science/index.htm *Well-developed UK educational resources site focusing on ages 5–11 11/3/01*

ENVIRONMENT IN GENERAL

http://www.agu.org/sci_soc/sci_soc.html   *The American Geophysical Union web site with good, basic information about land, water, space and air. It focuses on society and science. For example, there is, amongst many, a detailed article reviewing the flooding of the Mississippi 22/01/01*

http://info.tve.org/index.cfm   *TV trust for the environment concerns media representation of green issues. For example, a section in 'Last plant standing' called 'Treasure of the Andes' offers basic information on potatoes at http://info.tve.org/lps/doc.cfm?aid=649 22/02/01* http://www.worldbank.org   *The World Bank Group site. There are many useful links of environmental relevance here. (Go to the site map) 25/02/01*

http://www.snf.se/verksamhet/internationellt/index.cfm   *Swedish nature protection site 11/3/00*

ENVIRONMENTAL ACTION

Action is meaningless without knowledge. Many of the organisations in this section concentrate on politics and legislation. It would be useful if they could also provide scientific information on-line.

http://www.worldwildlife.org   *Worldwide Fund for Nature 23/01/01*

http://www.iclei.org   *International Council for Local Environmental Initiatives. Includes case studies on local environmental management at $3.50 each 23/01/01*

http://www.iisd.ca   *International institute for sustainable development. There is abundant information here on environmental management of, for example, the prairies, fisheries, mining, etc. The information is reliable and detailed, with lots of links 23/01/01*

http://www.greenpeace.org   *The most obvious information contains action and politics but a little digging reveals some concrete information, for example a history of ocean dumping, though legislation and government actions remain the understandable foci 23/01/01*

http://www.foe.co.uk   *Friends of the Earth homepage. Specialist information has to be paid for, though there are some free pamphlets. Not much available on-line; what is available tends to focus on campaign organisation 23/01/01*

http://www.unglobalcompact.org   *Global Compact. This is an official site of the United Nations. The front page took time to download but there is worthwhile information on almost every country in the world. A search for 'UK sustainable development' took me to a key government document and then many other links 23/01/01*

http://www.gn.apc.org   *Greennet – this site focuses on communications in relation to environment, peace, human rights and development. There is not much hard data here 23/01/01*

http://www.la21-uk.org.uk   *UK government site which offers specific details of local Agenda 21 activities. For example, the town of Frodsham in the UK had a credit balance of £500 on its LA21 activities 23/01/01*

http://www.igc.org/igc/gateway/enindex.html *Site for action and infor-mation on ecological issues. Up-to-date opinions but there are few hard data 8/2/02*

http://www.pbs.org/weta/planet/resource/comm.html *Planet Neighbor-hood is a not-for-profit US organisation which offers links to lots of other environmental action groups 5/3/00*

http://www.btcv.org *The British Trust for Conservation Volunteers. This is one of the ways you can get practically involved in the UK 2/2/01*

ENVIRONMENTAL EDUCATION

http://www.worldbank.org/html/edi/toronto/post/ra-eet96/tsld016.htm *Slide presentation on environmental education through agriculture training 25/02/01*

http://www.monkton.reserve.btinternet.co.uk/links.htm *Site maintained by a nature reserve in southern UK. Interesting with lots of links to sources of solid information 11/3/01*

http://www.naturegrid.org.uk/ng-html/index.html *Site associated with an environmental centre in Canterbury, England 5/3/00*

ETHICS AND SOCIAL ISSUES

http://www.sirc.org *Social Issues Research Centre. This is an interesting site which includes consideration of biological issues, such as essays on psychology of smell, and Desmond Morris on diet fads 23/01/01*

http://www.secondnature.org/programs/starfish/biblio.nsf *'Second Nature' is a non-profit organisation which gives useful reading lists for particular topics, e.g. hazardous materials and pollutants 23/01/01*

EVOLUTION

http://bioinfo.med.utoronto.ca/~lamoran/Evolution_home.shtml *The Uni-versity of Toronto gives links and content descriptions for a wide range of evolution sites, including those concerned with creationism, accompanied by appropriate warnings 1/4/01*

http://www.athro.com/evo/evframe.html *Commercial site but it comprises a good introduction to evolution 1/4/01* http://www.sigmaxi.org/amsci/articles/95articles/cdeduve.html *Essay on the origin of life by a recognised expert 1/4/01*

http://arnica.csustan.edu/biol1010/origins/origins.htm *Clear review of origin of life in university course notes 1/4/01*

http://tabla.geo.ucalgary.ca/~macrae/Burgess_Shale *Burgess Shale fossils described by the University of Calgary in Canada 1/4/01*

http://www.geo.ucalgary.ca/~macrae/talk_origins.html *This is a valuable evolution web site with a newsgroup based at the University of Calgary 1/4/01*

http://www.bbc.co.uk/education/darwin/exfiles *The Extinction Files, offered by the BBC 1/4/01*

http://eps.harvard.edu/people/faculty/hoffman/snowball_paper.html *On-line article from Harvard about Snowball Earth. Could it happen again? 1/4/01*

http://taggart.glg.msu.edu/paleo/paleo1.htm *Useful and well-illustrated pages from the University of Michigan 1/4/01*

FISHERIES

http://www.asf.ca/Overall/whoisASF.htm *The Atlantic Salmon Federation, international, non-profit, promotes the conservation of the wild Atlantic salmon and its environment 3/6/00*

FOREST ECOLOGY – SEE TREES AND FORESTS

FUNGI

http://www.aspergillus.man.ac.uk *Fungal Research Trust charity web site. They ask you to register (free) 30/3/01*

http://www.wisc.edu/botany/fungi/volkmyco.html *The engaging homepage of an American academic fungophile, leading on to many other useful fungus pages 30/3/01*

GENERAL BIOLOGY

http://www.biology.arizona.edu *On-line biology from the University of Arizona 24/3/01* http://phylogeny.arizona.edu/tree/phylogeny.html *The Tree of Life project hosted by the University of Arizona is an excellent source of biological data and images. It is a well-organised, easy-to-navigate, comprehensive and interesting review of the biological kingdoms. Every biologist should pay a visit 10/3/01*

http://www.microscopy-uk.org.uk/schools/mainscol.html#plant *Excellent site for images of plants, mammals and other biological materials with good supporting information 9/11/00*

http://biodidac.bio.uottawa.ca *This is an excellent site based at the University of Ottawa with a long list of images, arranged in taxonomic order 22/01/01*

http://www.eecs.umich.edu/mathscience/funexperiments/agesubject/ biology.html *Interesting high-school (equivalent to GCSE) lesson topics with practicals 7/03/01*

http://wise.berkeley.edu *School students and teachers might be interested to visit this site (Fig. 17.4) which offers a wide range of materials including text, images, practicals, discussion groups in attractive and stimulating formats 10/3/01*

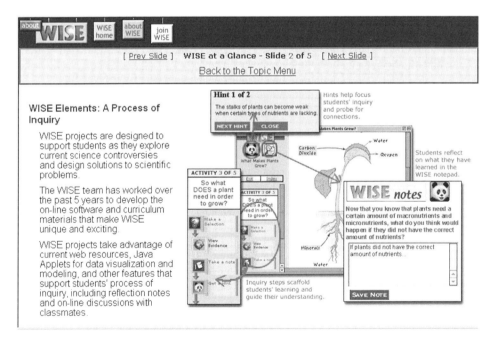

Figure 17.4 The WISE teaching and learning site at http://wise.berkeley.edu hosted by the University of California, Berkeley in the USA

GENETICS

http://biologylab.awlonline.com   *$30 buys you access to a suite of 12 web-based lab exercises on mitochondria, evolution, fruit fly genetics, etc. Worth the money if the specific program is what you need 23/01/01*

http://vector.cshl.org/dnaftb/41/concept   *Illustrated, easy introduction to genetics, including biographies from the Cold Spring Harbor Labs in the USA 10/3/01*

http://www-personal.ksu.edu/~bethmont/mutdes.html   *University of Kansas lecture notes on mutation, mutagens and DNA repair. Useful but unfortunately no illustrations 30/3/01*

www.marshmed.org/genetics/contents.htm   *Lots of solid information on human genetics is offered by the Center for Medical Genetics, Wisconsin, USA 30/3/01*

http://www.ornl.gov/hgmis/project/about.html   *Educational resources relating to the Human Genome Project. Extremely useful but from the appearance of this site, it was something done entirely by the USA. Europe had a hand in it too! Unfortunately, the UK version of this site at http://www.hgmp.mrc.ac.uk seems to pay attention only to the needs of those who are researchers in the subject. Don't the UK taxpayers deserve to get some interpretation of where their money has gone? 30/3/01*

http://www.dsmz.de/mutz/mutzdnaf.htm   *Isn't it a shame that though DNA fingerprinting was invented in England, there does not appear to be a UK web site which clearly and simply says what it is? Most web references to DNA fingerprinting refer to sites in the USA. For a change, this is a German one, mostly presented in English 30/3/01*

GOVERNMENTAL AND INTER-GOVERNMENTAL ORGANISATIONS

Governments

http://www.ukonline.gov.uk/online/ukonline/home   *Official starting point for all UK government web information 10/3/01*

www.gksoft.com/govt/en   *Governments on the web, presented by a consultancy. There are useful links here. For example, I went to the Belgian Institute of Natural Sciences and visited their research site on Arctic crustacean amphipods at http://www.natuurwetenschappen.be/ amphi/carcimage.htm (Fig. 17.5) 22/2/01, 17/7/01*

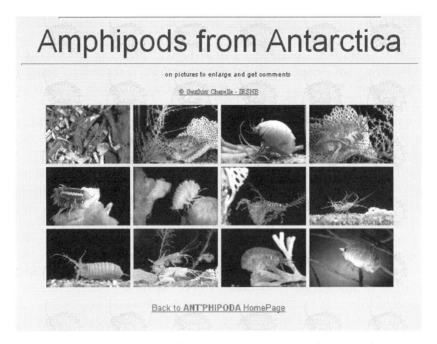

Figure 17.5 Belgian Institute of Natural Sciences research site at http://www.natuurwetenschappen.be/amphi/carcimage.htm

http://www.detr.gov.uk *Department of Environment, Transport and the Regions. Excellent site for detailed information on the UK environment. See the pdf file 'The environment in your pocket'. See .jpg files images>detibo0k and others 2/02/01*

http://www.dfid.gov.uk *Department for International Development Good search engine allows you to search government speeches and documents back to May 1997. Some of these documents contain concrete information, as well as policies 2/02/01*

http://www.doh.gov.uk/dhhome.htm *Department of Health. Concrete information on anything other than policy and management is difficult to find here. However, a poke round in the Research Findings Register showed that research indicates that older people are less likely to get access to essential diagnostic testing, which is not comforting for my 82-year-old father 2/02/01 If you would like to follow this up see the next web site*

http://tap.ccta.gov.uk/doh/refr_web.nsf/df64371528349537802568be0034e703/3b049f7bcd21ff70002569e70030d227?OpenDocument

http://www.environnement.gouv.fr *Ministry of the Environment in France. There is an English translation 3/6/00*

UNDP

http://www.undp.org *United Nations Development Programme site which mainly concerns money and projects but at http://www.unv.org/unvols/index.htm explains how you can become a UN volunteer 22/02/01*

UNED

http://www.unedforum.org *United Nations Environment and Development Forum (Fig. 17.6). This site is mainly economic and political but*

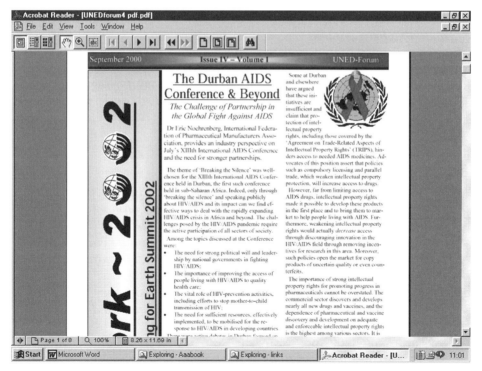

Figure 17.6 UNED-based newsletter called *UNED Forum* available on the web at http://www.unedforum.org (*Ref. Network 2002. T.Middleton, Ed. UNED Forum*)

there is some scientific information such as pdf file on 'Perspectives on freshwater - issues and recommendations'. This web site could be a useful starting point for general environmental issues 22/02/01

UNEP

http://www.unep.ch   *UNEP Geneva homepage of the United Nations Environment Programme. Extremely valuable site with a host of links. For example, I went to the Coral Reef Unit, which took me to links to a downloadable pdf atlas of East African Coastal Resources; investigation of Fisheries and Agriculture told me that only 7% of Kenya's fish harvest is marine 23/01/01*

http://www.unep.org/unep/elemlist.htm   *United Nations Environment Programme list of aspects to its programme. Lots of productive links concerned with environmental legislation here 23/01/01*

http://www.unep.org   *UNEP Nairobi includes Geo2000 - updated information on global environmental trends. Lots of good data are available here 2/02/01*

http://www.unep.org/unep/eia/eis   *Background to the UNEP net architecture 2/02/01*

http://www.unep-wcmc.org   *World Conservation Monitoring Centre is a good starting point for conservation links 22/02/01*

UNICEF

http://www.unicef.org/reseval   *Basic health data, country by country. Useful. Lots of information such as pdf files on 'Countdown to polio eradication'. Can be used as a starting point 2/02/01*

http://www.undp.org/seed/gef.html   *Biodiversity, climate change, international waters, ozone depletion, e.g. $0.851m for Azerbaijan to phase out noxious refrigerants 2/02/01*

www.earthsummit2002.org/roadmap/default.htm   *Roadmap to Earth Summit - annual representation of progress on sustainable development 2/02/01*

www.earthsummit2002.org/freshwater   *NGOs and freshwater. (Unfortunately there are problem links at the bottom of a paper by Jim Lamb, Severn Trent Water Authority. The links look useful but only go back to the head of the page) 2/02/01*

World Bank

http://www.worldbank.org/html/edi/toronto/post/sites.htm *World Bank list of official web sites relating to environmental education and training 25/02/01*

http://www.worldbank.org/html/edi/toronto/post/titoc.htm *Paper from an FAO consultant about getting sustainability awareness down to the grassroots. It illustrates the kind of information which can be obtained from the large inter-governmental agencies 25/02/01*

World Health Organization

http://www.who.int/home/map_ht.html *Useful. Under 'non-communicable diseases', Cambodia has 335 000 thalassaemia heterozygotes 2/02/01*

Miscellaneous political, industrial and and voluntary groups

Be wary. The information is mostly accurate but particular views are held by these organisations. Be particularly cautious about articles which don't have a clear author.

http://members.aol.com/seraoffice/index.htm *Labour environment campaign. This site offers a political dimension to some green issues – for example a paper on renewable energy – manufacturing in the knowledge economy at http://members.aol.com/seraoffice/ Brfrenew.htm 22/02/01*

http://www.igc.org/igc/gateway/index.html *Institute for Global Communications is a social justice web site based in California, and has links to anti-racism web sites etc. There is useful information here (see for example 'Biotech – the basics' by Rachel Massey. However, be wary, since a particular viewpoint is being expressed) 2/02/01*

http://www.iisd.ca/linkages *Resources for policy makers, listed by International Institute for Sustainable Development. There are mostly conference reports which need Realtime player to be downloaded, but some .txt files are available. For example, in the transcription of the 12th meeting of parties to the Montreal protocol, I find that Canada has notified a new ozone-depleting substance called hexachlorobutadiene . . . and it is the fourth new ODS to recently appear on the market 2/02/01*

http://www.bond.org.uk *BOND – network of voluntary-based organisations. One of the FAQs to which it offers an answer is 'How do I find work/volunteer jobs in the field of international development?' 2/02/01*

http://www.wbcsd.org *World Business Council for Sustainable Development. This site offers a particular industry-centred viewpoint but does point to useful information such as an independent report on a sustainable policy for paper (purchase price £30) 22/02/01*

http://www.sustdev.org/industry.news/072001/13.01.shtml *Electronic newsletter from Sustainable Development International (Fig. 17.7); for example, there is a note that 3000 scientists involved in the IPCC have stated that global warming is happening faster than expected. Although there is no specific attribution for this, the site has lots of links and attributable information 12/11/00*

Figure 17.7 On-line electronic newsletter from Sustainable Development International at http://www.sustdev.org/industry.news/072001/13.01.shtml

HISTOLOGY

http://www.usc.edu/hsc/dental/ghisto/index-alpha.html *Digital atlas of histology images presented by a US school of dentistry 18/3/01*
http://www.pathguy.com/histo/000.htm *Extensive range of histological images from the University of Health Sciences, Kansas City 1/4/01*

HISTORY AND PHILOSOPHY OF SCIENCE

http://www.hps.cam.ac.uk *Department of History and Philosophy of Science, University of Cambridge. Interesting links under 'Primary sources', e.g. X-ray image of Mrs Röntgen's hand 22.01.01*
http://www.people.virginia.edu/~rjh9u/sectip.html *This site is based on an unfortunately now-defunct course at the University of Virginia. There is a lot of solid information on subjects ranging from enzyme kinetics to many aspects of human genetics. It's a real shame it has stopped 30/3/01*

HORTICULTURE

http://aggie-horticulture.tamu.edu/tamuhort.html *Beautiful photos of plants plus good quality USA-based information 9/11/00*

IMMUNOLOGY

http://www.accessexcellence.com/AB/GG/antibodies.html *Illustrations of the basic principles 1/4/01*
http://immunology.org *Web site of the British Society for Immunology. De rigueur if you want to develop a career in UK immunology but there is no obvious free information for undergraduates 19.7.01*

JOURNALS (ACADEMIC)

It would obviously impossible to do justice to the journals. These are included merely to give a flavour.

http://journals.bmn.com/journals *This is probably one of the most important sites for the aspiring undergraduate and postgraduate scientist. It includes access to the contents lists and abstracts from an enormous list of journals and has sophisticated searching options. On-line purchasing of full texts is possible and some full texts are available free of charge 3/7/00*

http://www.pubs.royalsoc.ac.uk/proc_bio/proc_bio.html *This is the archive of the Proceedings of the Royal Society in the UK. Here you can get abstracts of high-quality articles on a range of biological subjects. The full-text articles can be purchased. On the other hand, having found the abstract here, you could always order a copy via your university library inter-library loan service 11/5/01*

http://www.idealibrary.com/servlet/useragent?func = showHome *You can get abstracts of articles from an enormous number of scientific journals via the Ideal site. However, there is only limited access to the full-text pdf files 1/4/01*

http://www.is.bham.ac.uk/resources/ejournals/e.htm *An A–Z of electronic journals, with links 5/3/00*

http://www3.interscience.wiley.com/cgi-bin/abstract/34206/START *Abstracts from Wiley Interscience Bioessays. Subscriptions are needed to see the pdf files. However, there is a good on-site search engine which means you could get a good range of general information to support an essay or practical write-up. The target abstract here is a useful introduction to the origin of the Eukaryota 20/12/00*

http://www.pnas.org/searchall/noframes.dtl *Search engine for a large number of journals, particularly relating to medical subjects. However, using the key words 'optimal foraging' several references to ecological journals were shown. Abstracts and full texts can be obtained, depending on the journals concerned 25/02/01*

http://www.sciam.com *Homepage of Scientific American. Some seren-dipitously useful articles are available. A search for 'Burgess' linked to a review of a book by Simon Conway Morris on the Burgess Shales and introduced an academic disagreement between Conway Morris and Stephen Jay Gould 25/02/01*

http://jxb.oupjournals.org/cgi/content/abstract/50/339/1541 *Journal of Experimental Botany. There is a lot of more advanced material here, available as abstracts or full-text pdf files 3/2/01*

MAPS

http://www.lib.utexas.edu/Libs/PCL/Map_collection/europe/Europe_ref 802637_1999.jpg *Useful map of Europe and access to other maps, though some are badly out-of-date 18/7/00* http://edcdaac.usgs.gov/glcc/glcc.html *Global land-cover characterisation by Landsat from NASA 18/7/00*

http://edcdaac.usgs.gov/gtopo30/gtopo30.html *United States Eros Geological Center offers access to beautiful topographical maps of the world 18/7/00*

http://SunSITE.Informatik.RWTH-Aachen.DE/Maps *Perry-Castañeda Library Map Collection at University of Texas at Austin 8/11/00*

MARINE SCIENCES

http://cbr.nc.us.mensa.org/homepages/BMLSS *Homepage of the British Marine Life Association. Generally not an aesthetically pleasing site but it leads to lots of information 25/02/01*

http://ourworld.compuserve.com/homepages/BMLSS/glaucus.htm *The journal of the British Marine Life Association. Some articles are printed in full. For example there is a useful introduction to oceanography by Amanda Young 25/02/01*

http://cbr.nc.us.mensa.org/homepages/BMLSS/Torpedo2.htm *Torpedo is the news service and Bulletin of the British Marine Life Association Includes lots of marine links and interesting images – news of meetings etc. 2/11/00*

http://ourworld.compuserve/homepages/BMLSS/EFora.htm *eFora is a web site which links to a wide range of other web sites on marine life 25/02/01–20/09/01*

http://ourworld.compuserve.com/homepages/BMLSS/Limpets.htm *British Marine Life Study Society 2/11/00*

http://cbr.nc.us.mensa.org/homepages/BMLSS/dogwhelk.htm *British Marine Life Society – discusses dog whelks 2/11/00*

http://www.hwdt.org *Hebridean Whale and Dolphin Trust. Good-quality information including photos, fact sheets, projects 9/11/00*

http://nhmi.org *Marine science education centre in the Florida Keys 25/02/01*

http://www.fao.org/fi/asfa/asfa.asp *FAO fisheries information site offering lots of concrete information 10/3/01*

http://www.pml.ac.uk/pml/pml_menu.htm *Homepage of the Plymouth Marine Laboratory, UK, one of the foremost research institutions 10/3/01*

http://seagrant.gso.uri.edu/riseagrant/factsheets/factsheets.html *Useful fact sheets from the Seagrant project hosted by the University of Rhode Island 1/4/01*

MEDICINE

http://www.bmj.com/cgi/content/full/321/7266/918/b   *British   Medical Journal. This highly esteemed journal offers an archive of some material published in the paper versions. An archive search with the keyword 'thyroid' yielded full-text articles which did not need a pdf reader 11/5/01*

http://www.foodstandards.gov.uk/sitemap.htm   *Site map of the UK Food Standards Agency 1/11/00*

http://www.ncbi.nlm.nih.gov/htbin-post/Entrez/query?uid = 9256411& form = 6&db = mb *Search engine for the USA National Library of Medicine 25/02/01*

http://www.mrc.ac.uk   *Medical Research Council – this is the official homepage. Judicious searching can lead to some useful hard information published by the Department of Health 25/02/01*

http://www.doh.gov.uk/smac1.htm   *MRC official report on antibiotic drug resistance as pdf file (24 tables and 16 figures!) 25/02/01*

http://www.mcri.ac.uk/default.html   *Marie   Curie   Cancer   Research Institute. Irritating molecule as a navigation tool but click on site map to see your way around. By following up individual research groups, I was able to access a full-text version of a paper on general transcription initiation factor TFIID published in Proceedings of the National Academy of Sciences 25/02/01*

http://www.icnet.uk *Imperial Cancer Research Fund 25/02/01*

http://www.crc.org.uk   *Cancer Research Campaign – some basic information (e.g. statistics) is available at http://www.crc.org.uk/press/ press_office_intro.html on this site 25/02/01*

http://dspace.dial.pipex.com/lrf-//   *Leukaemia research fund. Very little hard data here 16/10/00*

http://www.med.ic.ac.uk/ludwig/home.htm   *Ludwig   Inst.   Cancer Research. As with the other research groups, I can follow a line, for example to the tumour suppression gene group at http://www.med. ic.ac.uk/ludwig/xingroup.htm, where I find that 'an Rb-mdm2-p53 trimeric complex is active in p53 mediated transrepression' 25/02/01*

http://www.mhsource.com/narsad   *National Alliance of Research on Schizophrenia and Depression. A search in the site archive tells me that women with seasonal affective disorder or bulimia are more likely to have a particular polymorphism in a serotonin gene called tryptophan hydroxylase (TPH) 25/02/01*

http://www.schizophrenia.com   *Information for patients and families including links to authoritative articles 25/02/01*

http://www.aspirin.org   *Aspirin Foundation of USA. Site includes news and documentation 9/11/00, 25/02/01*

http://www.hiv-druginteractions.org   *HIV–drug interactions site hosted by the University of Liverpool. Hard data are available, for example a poster on measurement of nucleoside analogue triphosphates by enzymatic assay in HIV patients in a clinical setting 25/02/01*

http://www.bsereview.org.uk   *BSE Controls Review, is a site presented by the Food Standards Agency of the UK. Lots of BSE information is presented here 1/11/00*

http://www.ama-assn.org/ethic/vm/healing/heal0500.htm   *Images of Healing and Learning is a stimulating site maintained by the American Medical Association; for example it uses images of the American Civil War to examine the role of doctors in war 11/3/01*

http://www.ama-assn.org/ama/pub/category/2704.html   *Site map of the American Medical Association 1/11/00*

http://www.phls.co.uk/facts/index.htm   *Fact sheets on disease from the Public Health Laboratory Service of the UK 22/3/01*

http://www.msf.org   *Médecins sans Frontières is an international aid organisation which began in France. It now has offices in 18 countries. This is their homepage 5/3/00–20/9/01*

http://www.who.int/tdr   *WHO special programme for research and training for tropical diseases 5/3/00, 24/3/01*

http://www.rcpath.org/activities/list.html   *Web site of the Royal College of Pathologists which has lots of reliable and useful biomedical information. For example, it was possible to obtain a full-text pdf government publication on 'Revised Advice on Laboratory Containment Measures for work with Tissue Samples in Clinical Cytogenetics Laboratories – HIV and hepatitis' 17/7/01*

http://www.embase.com/home.isa   *EMBASE gives access to high-quality, current, biomedical and pharmacological information. You can search more than 13 million bibliographic records with abstracts. However, to get access you need a subscription key code from your university librarian 19/7/01*

MICROBIOLOGY

http://www.phls.co.uk   *Public Health Laboratory Service of the UK. There is lots of free, reliable information here 19/7/01*

http://www.bact.wisc.edu/MicrotextBook/ClassAndPhylo/natural.html   *University of Wisconsin hosts a free web textbook of microbiology 1/4/01*

http://www.slic2.wsu.edu:82/hurlbert/micro101/pages/101hmpg.html
*Washington State University hosts a free web textbook of microbiology
1/4/01*

http://www.md.huji.ac.il/microbiology/book/toc.htm   *The Hebrew University, Jerusalem hosts a free web textbook of microbiology 1/4/01*

http://microbiol.org   *This is the Microbiology Network homepage and is
probably the starting point for mainly postgraduate microbiology 1/4/01*

http://people.ku.edu/~jbrown/bugs.html   *'Bugs in the news' hosted by
Kansas University. This site makes microbiology interesting 17/3/01*

http://bugs.uah.ualberta.ca/webbug/index.htm   *Important site for microbiology on the web hosted by the University of Alberta. It clearly leads
to a wide range of sites 17/03/01*

http://dbs.ucdavis.edu/courses/w01/mic102l/files/041.html   *Lab   notes
from microbiology course at the University of California, Davis.
There are good links to other microbiology sites and lots of useful hard
data 25/02/01*

http://www.iah.bbsrc.ac.uk/virus/Picornaviridae/picornavirus.htm   *Picorna-
virus homepage hosted by UK government department. Lots of virology
links are here 10/3/01*

http://phylogeny.arizona.edu/tree/eubacteria/eubacteria.html   *Tree   of
Life site at the University of Arizona 10/3/01*

http://www.thebugstopshere.co.uk/invisible_world_goodies.shtml   *Useful
site for junior school teaching, introducing microbes. However, the
commercial logo of the sponsor is a little too prominent for my taste
17/3/01*

http://www.microbelibrary.org/FactSheet.asp?SubmissionID = 678   *American
Society for Microbiology curriculum resources offers, among many other
things, an undergraduate practical to investigate antibiotic action
17/3/01*

http://www.asmusa.org/mbrsrc/archive/SIGNIFICANT.htm   *Timeline for
history of microbiology from the American Society for Microbiology
17/3/01*

http://www.microscopy-uk.org.uk/mag/indexmag.html   *On-line magazine
hosted commercially but with lots of useful information and images
18/3/01*

http://www.across.utas.edu.au/program1.html   *Research in separation
sciences at three Australian universities. The introductions to the
research projects give useful information 18/3/01*

http://helios.bto.ed.ac.uk/bto/microbes/penicill.htm#   *The emergence of
antibiotic resistance. Very clear description of antibiotic drug
resistance from the University of Edinburgh 1/4/01*

MISCELLANY

http://europa.eu.int/eur-lex/en/map.html *Web site for the European Commission legislation. Here you can find out what legislation applies to the BSE and Foot and Mouth epidemics 1/4/01*

MOLECULAR BIOLOGY

http://www.wnet.org/innovation/index.html *The non-profit Public Broadcasting Service of the USA has put together any excellent site which offers easy explanations for the technical and other aspects of molecular biology, appropriate for undergraduates, A-level students and the interested citizen 30/3/01*

http://www.accessexcellence.org/AB/GG/#Anchor-From-14210 *Genentec have contributed to this site called 'Accessexcellence' which offers molecular biology and genetics information which is particularly appropriate for undergraduates 26/3/01*

http://www.sanger.ac.uk/sitemap/hgp.shtml *Sitemap for the human genome project. Even a non-specialist could find this an interesting site – they probably wouldn't understand much but it is aesthetically pleasing, and it concerns us all 10/3/01*

http://www.sciam.com/explorations/2001/021201humangenome *Well-linked Scientific American article which describes the Human Genome Project 25/02/01*

http://www.sciam.com/1999/0799issue/0799scicit4.html *Stem cells come of age. Well-linked article from Scientific American 25/02/01*

http://www.nhgri.nih.gov/educationkit *US National Institute of Health web site explaining the Human Genome Project 10/3/01*

http://vector.cshl.org/dnaftb *Introduction to molecular genetics from the Government Cold Spring Harbor Labs in the USA. Well illustrated and linked 10/3/01*

http://vector.cshl.org/dnaftb/15/concept *An excellent, free, web-based learning resource called the 'DNA Learning Center', hosted by the famous Cold Spring Harbor Laboratory in the USA 26/3/01*

http://www.tbi.univie.ac.at/~ivo/RNA *Free suite of software for RNA structure prediction and comparison 26/3/01*

http://www.iacr.bbsrc.ac.uk/notebook/index.html *An excellent, free, web-based learning resource called the 'Molecular Biology Network', hosted by a British Research Council 26/3/01*

http://www.biology.ucsd.edu/others/dsmith *Sequence   analysis   and molecular biology links hosted by the La Jolla Labs, University of California 26/3/01*

MUSEUMS

http://www.24hourmuseum.org.uk   *General search engine for museums in the UK 1/4/01*

http://www.nhm.ac.uk *Web site of the Natural History Museum of London offers many resources but a strangely limited search engine 1/11/00*

http://www.mnh.si.edu *The Smithsonian Museum of Natural History in Washington DC is one of the great museums of the world 1/11/00*

http://www.exploratorium.edu *On-line science museum 1/11/00*

http://www.skansen.se/eng/dvindex.htm *An   open-air   museum,   in Stockholm. There is information on Scandinavian animals, plants and agriculture 24/3/01*

http://www.rbge.org.uk *The Royal Botanic Gardens, Edinburgh 18/3/00*

http://www.neaq.org/learn/index.html *The New England Aquarium. There's lots of fishy stuff here 18/3/00*

http://sciencenorth.ca/whoweare/index.html *Science North – a bilingual science museum in Ontario. Visit this site and learn French (or English!) by studying science in two languages 18/3/00*

NEUROPHYSIOLOGY

http://www.sciam.com/2000/0400issue/0400tsien.html *Building   a brainier mouse. Scientific American featured article which includes many aspects of neurophysiology, well linked to a glossary 25/02/01*

http://www.st-and.ac.uk/~seeb/pheromone/index.html *Odour in mate recognition. Royal Society exhibition presentation 9/11/00, 25/02/01*

http://www.hhmi.org/senses/d/d110.htm *Howard   Hughes   Medical Institute Research on smell. Easily accessible information for undergraduates 10/3/01*

http://cvs.anu.edu.au/andy/beye/beyehome.html *See the world through the eyes of a bee at a university web site; the papers were developed by a PhD researcher 26/3/01*

http://www.augie.edu/perry/frames.htm *Virtual tour of the ear. Lots of solid, visual information is offered is offered by this college site 26/3/01*

http://retina.anatomy.upenn.edu/~lance/retina/retina.html   *University of Pennsylvania tells you everything you want to know about the retina 26/3/01*

NEWS AND REVIEW JOURNALS

Very academically rigorous tutors may not consider these journals an appropriate literature source. However, they can give first-year under-graduates (and others!) a good general picture of current issues. They offer many services, including job searches and are generally less depressing than newspapers.

http://www.newscientist.com   *Long-established   UK-based   science review magazine offering lots of facilities including a free newsletter and over 1600 web links with commentaries 3/7/00*

http://whyfiles.org/index.html   *Science behind the news from a US commercial page 1/4/01*

http://www.findarticles.com/cf_0/PI/index.jhtml   *This   search   engine finds articles in magazines and newspapers. It finds some useful information, passed on as full-page articles 1/4/01*

http://www.popsci.com   *Popular science magazine which includes some interesting articles 17/07/01*

http://www.the-scientist.com/homepage.htm   *Popular USA-based review science magazine with freely available on-line material. Monthly updates can be sent to your inbox 3/7/00*

http://www.ec.gc.ca/envirozine/english/home_e.cfm   *On-line   environ-mental magazine from Canada. Reasonable quality with some interesting things, e.g. at   http://www.ec.gc.ca/EnviroZine/english/ issues/08/feature2_e.cfm there is an article on mapping smog 27/7/01*

NUCLEAR RADIATION ISSUES – SEE RADIATION

ORNITHOLOGY

http://www.skof.se/eindex.html   *Web site of the Ornithological Society of Skania in southern Sweden, in English 11/3/01*

http://www.mna.hkr.se/~nec/havseagle.htm   *Technical information by a Swedish academic on the white-tailed fish eagle in Sweden 3/6/00*

http://www.the-owl-barn.com   *On-lines sales of bird goodies 11/3/01*

PALAEONTOLOGY

http://www.nhm.ac.uk/interactive/vrml/pictures.html   *Virtual   reality images of fossils at the London Natural History Museum, the main site   being   at   http://www.nhm.ac.uk/palaeontology/index.html 11/11/00*

http://www.ucmp.berkeley.edu/index.html   *Museum of Paleontology at the University of California, Berkeley 11/11/00*

http://www.monkton.reserve.btinternet.co.uk/fossils.htm   *Comprehensive description of the palaeontology of part of south-east England 11/3/01*

http://www.dnr.state.md.us/fisheries/education/horseshoe/horseshoefacts. html   *The horseshoe crab – A living fossil? 1/4/01*

PARASITOLOGY AND DISEASE

http://www.phls.co.uk/facts/index.htm   *Excellent UK government site with lots of information and links to other sites. For example, a click on 'bacterial drug resistance' takes you to graphs of real data under the heading 'Bacteraemia' and then to help interpret these data there are links to technical articles such as the British Medical Journal at http://www.bmj.com/cgi/content/abstract/320/7229/213 19/7/01*

http://www.mosquito.org *Information   on   mosquitoes   from   the American Mosquito Control Association, based at Rutgers University. Most information has to be purchased 25/02/01*

http://www.hortnet.co.nz/publications/hortfacts/hf401043.htm   *Informative site of the Horticultural and Food Institute of New Zealand. Good on insect pests 28/02/01*

http://nrs.ucop.edu/reserves/coal/ecorep.html   *Coal Oil Point Reserve: 1993–94 Research Reports. Here there is a report on research on the killifish, host to Euhaplorchis californiensis, a fluke 28/02/01*

http://www.life.sci.qut.edu.au/LIFESCI/darben/paramast.htm   *Useful and interesting images of animal parasites provided by Queensland University of Technology in Australia but broken links make this look like a dying site 28/02/01*

http://www.uni-hohenheim.de/~www380/parasite/start.htm   *Here   the University of Hohenheim in Germany provides masses of solid information on parasitic weeds 28/02/01*

http://www.science.siu.edu/parasitic-plants/index.html   *Interesting   site about parasitic plants with images links and data. For example, see the glossary at http://www.science.siu.edu/parasitic-plants/Terms.html 28/02/01*

http://www.cdc.gov/travel/mailinfo.htm *US centre for disease control and prevention, offering information on malaria amongst others 28/02/01*

http://ibs.uel.ac.uk/mosquito/biblio.htm *Mosquito bibliography (UK) 28/02/01*

http://www.uel.ac.uk/mosquito *European Mosquito Bulletin. Good information is offered here – for example an article entitled 'Aedes cretinus: is it a threat to the Mediterranean countries' 28/02/01*

http://www.fao.org/wairdocs/ilri/x5492e/x5492e04.htm *Article on epidemiology of parasites. Good basic information 2/11/00*

http://www.biologie.uni-erlangen.de/parasit *Good parasitological research web site from the German University of Erlangen-Nürnberg 16/11/00*

http://martin.parasitology.mcgill.ca/jimspage/hookworm.htm *Nice picture of a hookworm 28/02/01*

http://martin.parasitology.mcgill.ca/jimspage/BIOPAGE/outline.htm *Good information available at McGill University in Canada 28/02/01*

http://martin.parasitology.mcgill.ca/jimspage/namerica.htm *Nice picture of Necator americanus 28/02/01*

http://www2.umdnj.edu/eohssweb/bbp/intro.htm *Learning outcomes for a university-level self-study module on blood pathogens 1/4/01*

http://www.maff.gov.uk/animalh/diseases/fmd *British government web site (Department for Food, Environment and Rural Affairs) concerned with Foot and Mouth disease 1/4/01*

PESTS

http://www.denniskunkel.com *Homepage of Denis Kunkel, a professional zoological photographer. There are some excellent images here. Look at the diatoms 22/01/01*

http://www.ext.vt.edu/departments/entomology/factsheets/mealworm.html *One of a series of entomological fact sheets from the Virginia Polytechnic Institute in the USA 22/01/01*

http://www.biosci.ohio-state.edu/~parasite/periplaneta.html *Useful fact sheets from Ohio University, with images of a wide range of pests and parasites 1/4/01*

PHYSIOLOGY

Animal

http://www.csa.com/crw/websites.html *This chemoreception web site is a good starting point for any work in this area 9/11/00*

http://www.glencoe.com/sec/science/cgi-bin/splitwindow.cgi?top =
http://www.glencoe.com/sec/science/top2.html&link = http://sln.fi.edu/
biosci/heart.html *The Franklin Museum in the USA presents the biology
of the heart, with many images. This URL is very long. Find the page
yourself by searching at http://www.glencoe.com/sec/science 18/3/01*

http://ajpgi.physiology.org *The American Journal of Physiology gives
access to its abstracts 18/3/01*

http://lala.mpimf-heidelberg.mpg.de/~holmes/muscle/muscle1.html
Details of muscle contraction from a university web site 23/03/01

http://classes.aces.uiuc.edu/AnSci308/intro.html *Course on lactation
physiology from Urbana-Champaign College in Illinois 30/3/01*

http://www.culver.edu/homepages/faculty/jcoelho/biol309/coldmous.htm
*Culver-Stockton University offers lab practicals on respiration in
homeotherms 1/4/01*

http://www.arce.ukans.edu/book/eye/struct.htm *This University of Kansas
site describes the structure and function of the mammalian eye 1/4/01*

http://www.spb.wau.nl/mt/iawqstrproject *Applied biology – respiration
based control of sewage sludge management 1/4/01*

http://www.mhhe.com/biosci/genbio/maderbiology/supp/homeo.html
*Homeostasis described via link at the McGraw-Hill publisher's web site
22/07/01*

PLANT

http://www.microscopy-uk.org.uk/schools/images/stomata.html *Images
and descriptions of stomata 9/11/00*

POLLUTION

http://www.nrdc.org/health/default.asp *National Resources Defense
Council in the USA offers solid information on toxic and other forms
of pollution 5/3/00*

http://www.foe.co.uk/campaigns/industry_and_pollution/factorywatch
*Friends of the Earth site which allows you to get details of local
pollution by postcode 1/11/00*

http://www.homecheck.co.uk *Information on landfills, landslip, air
quality, etc. in the UK are available by postcode from this site 1/11/00*

http://www.scorecard.org *US site with comprehensive data on the
health risks of pollution in various parts of the USA 1/11/00*

http://www.ser.org   *Society for Ecological Restoration web site. There is no raw information but there are useful abstracts from 1993, 1995, 1996 16/10/00*

http://www.enviroweb.org/edf/sitemap.html   *This site offers comprehensive global warming statistics – some download problems, so use the sitemap 9/11/00*

http://www.epa.gov   *Main web page for the US Environmental Protection Agency 10/3/01*

http://www.epa.gov/globalwarming   *EPA–USA govt air pollution site 9/11/00*

http://www.aeat.co.uk/netcen/airqual/home.html   *Government department – air pollution forecasting in the UK 9/11/00*

http://www.environment.detr.gov.uk/airq/airpoll/index.htm   *Government department – background information on air pollution 9/11/00*

http://www.cce.cornell.edu/aquaticinvaders   *National aquatic nuisance species clearing-house, e.g. zebra mussels 9/11/00*

http://www.sepa.org.uk   *Web pages of the Scottish Environmental Protection Agency which offers lots of technical information including reports on Bathing Water, State of the Environment, National Waste Strategy, Habitat Enhancement Initiative, Radioactivity in Food and the Environment 10/3/01*

http://www.env.gov.bc.ca/epd   *Pollution prevention and remediation in Canada is described at this Canadian government site 1/4/01*

http://www.envir.ee/baltics/frame1.htm   *Heavy metal pollution in the Baltic. There's lots more on pollution here 1/4/01*

PUBLIC UNDERSTANDING OF SCIENCE

http://www.wellcome.ac.uk/en/1/mismiscnesos.html   *Review of PUS by esteemed research foundation 17/7/01*

RADIATION

http://europa.eu.int/comm/environment/radprot/index.htm   *The official European radiation protection site which directs you to information such as a 'Pilot on the radiological exposure of the European Community from radioactivity in North European marine waters'. It also directs you to legislation at http://europa.eu.int/eur-lex/en/lif/ind/en_analytical_index_15.html 17/07/01*

http://www.pubs.royalsoc.ac.uk/proc_bio/proc_bio.html   *Abstract of article*

*published in the Royal Society journal on high mutation rate in offspring
of Chernobyl accident liquidators*

http://www.bmj.com/cgi/content/full/321/7266/918/b   *British   Medical
Journal article on newborns affected by Chernobyl 11/5/01*

http://www.llrc.org/index.html   *Index page of the Low Level Radiation
Campaign. The site offers full-text versions of articles taken from other
journals 11/5/01*

http://www.uvm.edu/~radsafe/newsletter/biological.effects.50.html   *Introduction to radiation biology from the University of Vermont in the USA
1/4/01*

http://www.arpansa.gov.au/new_fct3.htm   *Dose limits fact sheet from
the Australian radiation protection authority 1/4/01*

http://tis-nt.eh.doe.gov/ohre   *US Dept of the Environment web page
concerning human radiation experiments 10/3/01*

http://cnts.wpi.edu/rsh   *Site concerning radiation, science and health
hosted by Worcester Polytechnic Institute. This is a good starting place
for anything which involves ionising radiations 10/3/01*

http://cnts.wpi.edu/RSH/Docs/Other%20Docs/UNSCEAR%202000/index.html
*2000 Report to the General Assembly of the UN, on radiation sources
and health, esp. Chernobyl 1/10/01*

http://www.ssi.se/english/index.htm   *Swedish   Radiation   Protection
Institute site which provides comprehensive information in English
1/12/00*

http://www.badc.rl.ac.uk   *British Atmospheric Data Centre. NERC and
other information. Login needed for some access. 1.5 million files of
data from satellites, balloons, aircraft computer models. Hosted by
Rutherford Appleton Laboratories 12.1.01*

http://www.sepa.org.uk/publications/consultations/tritium_in_scotland.pdf
*Report from Scottish Environmental Protection Agency on tritium
contamination in Scotland 7/3/01*

http://www.open.gov.uk/rwmac/reprocess/01.htm   *Radioactive   Waste
Management Advisory Committee which supplies information in, for
example, the waste implications of reprocessing 24/11/00*

http://www.bnes.com/nucwaste.htm   *British Nuclear Energy Society
report on radioactive waste management 1/12/00*

http://www.hse.gov.uk/press/e00177.htm   *Managing radiation risk –
information form the UK Health and Safety Inspectorate concerning
industrial radiography 1/12/00*

http://www.ntp.org.uk/rpda   *List of abstracts of articles from Nuclear
Technology Publishing. For example, there is an abstract on excessive
pharyngeal radiation 1/12/00*

http://ccnr.org/radiation_standards.html   *A critical review, from Canada, of ICRP exposure standards 1/4/01*

http://www.congrex.com/valdor   *Nuclear risk assessment conference – 'Value in decisions on risk', Sweden (triennial) 1/12/00*

http://www.irpa.net   *International Radiation Protection Association. Some pdf documents available 5/1/00*

http://www.srp-uk.org   *Web site of the Society for Radiological Protection. There is access to lots of fundamental information such as a synopsis of the Doll Report on cancer risks associated with electromagnetic radiations at http://www.srp-uk.org/doll.html#Group 7/3/01*

http://www.ieer.org/ensec/no-4/no4frnch/protecti.html   *A French perspective on ionising radiation protection – in French 1/4/01*

http://www.multimania.com/mat66/faits_deran_tchernobyl.html#anchor 228652   *Birth defects after Chernobyl. Technical magazine article in French 1/4/01*

SCIENCE IN GENERAL

http://www.bbc.co.uk/science/scienceshack   *Adam Hart-Davies runs a helpdesk for scientific questions, and suggests home experiments. Good serendipitous site. My last visit explained Foot and Mouth and explained why organisms yawn 1/02/01*

SOCIETIES – ACADEMIC

There are thousands of academic societies and journals. All that can be done here is to use a few to show how concrete information can be obtained.

http://www.royalsoc.ac.uk/news/index.html   *The Royal Society is one of the foremost scientific societies in the world. Although its publications have to be bought, journal abstracts are available at http://www.pubs. royalsoc.ac.uk/proc_bio/proc_bio.html and a series of news briefings could be useful for undergraduate assignments – for example 'Why aren't we all bacteria' 1/11/00*

http://www.iob.org   *Institute of Biology – current but not past articles can be accessed 9/11/00*

http://www.ase.org.uk   *Web site of the Association of Science Education (UK). Curiously, under 'projects', a keyword search of the science data-*

base using either 'transpiration' or 'osmosis' gave no hits. Searching by 'Unit of Study' revealed links, for example, to teaching resources at the University of Kansas in the USA where it was possible to find a lesson plan (without images) on transpiration, at http://explorer.scrtec.org/ explorer/explorer-db/rsrc/783750870-447DED81.1.htm 9/11/00

http://www.systass.org   *This Systematics Association (UK) site is rather typical of a society's site in that it says what it does, announces conferences, and points the visitor to literature that can be bought, but does not make available any concrete data 9/11/00*

SOCIETY AND BIOLOGY

http://www.wsu.edu:8001/vwsu/gened/learn-modules/lmindex.html *Washington State University offers modules on world civilisation which include solid data – for example on soils at http://www. wsu.edu:8001/vwsu/gened/learn-modules/top_agrev/2-soil/soil2.html 1/11/00*

SOIL BIOLOGY

http://classes.css.wsu.edu/css431   *Web page of a soil biology class at Washington State University which offers useful links to a number of soil-related web sites; for example, there are several on 'worms' 10/3/01*

SUSTAINABILITY

http://www.cat.org.uk/catpubs/tipsheet.tmpl?sku = 03   *Centre for Alternative Technology fact sheets – this one is on reed-bed sewage treatment 1/4/01*

http://www.balticnet.de/Documents/Ressourcenmanagement.html   *Scroll down this German web page. There is a huge list of web sites related to sustainability, not just on the Baltic 1/4/01*

TAXONOMY

http://www.nhm.ac.uk/zoology/taxinf/index5.html   *Darwin Initiative project at the Natural History Museum of London, investigating the use of the internet for taxonomy, using polychaete annelids as a model 10/3/01*

http://www.nhm.ac.uk/science/projects/fff/index.htm *Details of classification and types of plants from the Natural History Museum. Put in your postcode to get a species checklist and then you can see images and details of plants which live near you. Excellent 10.3.01*

http://www.pml.ac.uk/pml/Taxonomy.html#Current *This web site describes the taxonomy on the internet programme of the Plymouth Marine Laboratory, UK 10/3/01*

http://manandmollusc.net/taxonomy.html *Introduction to taxonomy using the molluscs as an example. Any students who are not sure what taxonomy is should visit this site 3/11/00*

http://www.rmetzner-greenearth.org/geo_kingdoms.html *Synopsis, by Lyn Margulis, of the five-kingdom classification 30/3/01*

http://www.ucmp.berkeley.edu/clad/clad4.html *The University of California Museum of Paleontology at Berkeley offers an introduction to phylogenetic systematics 6/6/00*

www.nrm.se/ev/musselnyckel/index.html.en *Key to freshwater mussels provided by Swedish Museum of Natural History 1/4/01*

TERMINOLOGY

http://www.mblab.gla.ac.uk/~julian/Dict.html *Dictionary of Cell Biology. You can pay for on-line access, or access for free 1 day in 90 10/3/01*

TREES AND FORESTS

http://www.british-trees.com *Aesthetically pleasing and informative site on all aspects of British trees 11/1/01*

http://www.communityforest.org.uk *Information on the 12 community forests run by the Countryside Agency/Forestry Commission. There is lots of solid information here – for example a paper on wood as an alternative source of energy at http://www.communityforest.org.uk/swcb.html 10/3/01*

http://www.forestry.gov.uk *The Forestry Commission web site. Includes an interactive question and answer section. Research projects are listed but little information is directly available 10/3/01*

http://www.rfs.org.uk   *Royal Forestry Society including contents pages of Quarterly Journal of Forestry but no access to abstracts 10/3/01*

http://www.totap.org.uk *Trees of Time and Place, supporting collection and planting of tree seed. Site maps showing planting are available 10/3/01*

http://www.un.org/esa/sustdev/forests.htm   *United Nations Sustainable Development Forum focused on forests of the world. For example, specific information on Russia is available at http://www.un.org/esa/agenda21/natlinfo/countr/russia/natur.htm#forests 10/3/01*

UNIVERSITIES

Many universities have been cited elsewhere in this list so only a few are mentioned here.

http://www.pbrc.hawaii.edu   *University of Hawaii. This site is typical of university sites whereby links to staff homepages reveal valuable information about their research, such as that offered at http://xenon.ctahr.hawaii.edu/Sun.htm regarding molecular biology applied to shrimp aquaculture 1/11/00*

http://www.dbs.ucdavis.edu   *Homepage of University of California at Davis 1/11/00*

http://www.colostate.edu/Depts/Biology   *University of Colorado biology site. Some snippets of information are available on the homepages of some staff members such as Shanna Carney at http://lamar.colostate.edu/~secarney 28/02/01*

VIROLOGY

http://www.virology.net/Big_Virology/BVHomePage.html   *'All the virology on the Net' is THE virology site comprising a wonderful symbiosis between virologists. Here is its Big Picture Book of Viruses 30/3/01*

ZOOLOGY (SYSTEMATIC)

http://mcnet.marietta.edu/~mcshaffd/invert   *Invertebrate zoology lecture notes as bulleted lists hosted by Marietta University in the USA 18/3/01*

http://www.ucmp.berkeley.edu/help/topic.html   *Masses of information on a host of marine organisms, organised by taxonomic group 1/4/01*

Porifera

http://animaldiversity.ummz.umich.edu/porifera.html   *Sponges   at   the University of Michigan 30/3/01*

Cnidaria

http://www.planet-pets.com/plnthydr.htm   *The   general   biology   of Hydra, with photos 30/3/01*

Platyhelminthes

http://www.cs.tufts.edu/~cabotsch/bulloughs/invertebrates/worms/ flatworm.html   *Flatworms at Tufts University, USA 30/3/01*

Molluscs – a self-indulgence

http://www.soest.hawaii.edu/tree/cephalopoda/cephalopoda.html#top
   University site which focuses on the cephalopods (octopus, squid, cuttlefish). Lovely images 1/11/00
http://www.cephbase.dal.ca *Life   history,   distribution,   catch   and taxonomic data on cephalopods 1/11/00*
http://www.mnh.si.edu/cephs   *Smithsonian Museum cephalopods 1/11/00*
http://www.abdn.ac.uk/eurosquid   *Comprehensive   and   well-linked   UK web site concerning world cephalopods, based at the University of Aberdeen 1/11/00*
http://peet.fmnh.org   *Field   Museum   of   Natural   History   in   the   USA – bivalve pages 1/1/00*
http://www.bigai.ne.jp   *This mollusc enthusiast site is both informative and anachronistic. Unfortunately for non-Japanese-speakers, lots of it is inaccessible. I hope the author will extend the English-language version. For example, there is an intriguing love story only tangentially related to molluscs but actually about the pollution of a beach. It begins: 'The wave coming in the beach of Tsuyazaki will go out soon, but our love have no way but going out of this world'. Irresistible 7/3/01*
http://manandmollusc.net   *Typical privately maintained site on man and molluscs with lots of solid information and good links 11/3/01*

Crustacea

http://www.nmnh.si.edu/iz/copepod *This is a typical museum site, based at the Smithsonian, which links to sites all over the world and opens up a huge body of information for enquiries into this taxonomic group 1/11/00*
http://www.fathom.com/story/story.jhtml?story_id = 121900 *Barnacles. A lecture delivered at the Natural History Museum on London 30/3/01*

Insecta

http://www.iacr.bbsrc.ac.uk/examine/aphids.html *Exploitation of aphid monitoring systems in Europe. Aphid control site with lots of zoological and agricultural information 1/11/00*
http://www.colostate.edu/Depts/Entomology/courses/en507/student_papers_ 1995/hawley.html *Student essay on aphids (part of a bank of such essays). Tutors should check for plagiarism against their own students' work 1/11/00*
http://www.nhm.ac.uk/entomology/index.html *Entomology at the Natural History Museum of London 10/3/01*
http://www.insects.org *Beautiful privately maintained site on insects with lots of solid information and images 11/3/01*
http://ucdnema.ucdavis.edu/imagemap/nemmap/ENT10/ent10syll.htm *A range of excellent support information is made available by the University of California, Davis 26/3/01*
http://animaldiversity.ummz.umich.edu/accounts/tenebrio/t._molitor$narra tive.html *Basic entomological information from the University of Michigan – in this case on the Flour Beetle 3/2/01*

GENERAL ASPECTS

Child safety on the internet

http://www.safekids.com/child_safety.htm *Excellent, official site of the National Center for Missing and Exploited Children in Arlington, Virginia, USA. There are checklists and lots of good advice here 3/4/01*

Serendipitous sites

http://www.indiana.edu/~pietsch/home.html *Eclectic source of links called Shufflebrain, hosted by Indiana University in the USA. Although it is biased towards neurobiology there is a lot of information here and it is worth a browse at coffee time 5/3/00*

http://www.hhmi.org/grants/lectures *Holiday lectures in science from the Howard Hughes Medical Institute in the USA 5/3/00*

Selecting a UK university at which you want to study

http://www.ucas.com/ucc *Information concerning access to higher education courses 7/3/01*

http://www.lueneburg.de/index.html *Typical web site for a continental European university, with English translation 3/3/00*

Careers

http://www.careers-portal.co.uk/?m = c_map.htm *On-line careers service 7/3/01*

http://www.jobsite.co.uk *Jobs. The keywords 'scientific' and 'biology' for the previous 7 days yielded 25 jobs, ranging from sales manager (£50 000) to biochemist technician (£17 000) to clinical trials associate (£20 000) 7/3/01*

http://www.virology.net/Articles/jobs.html *'All the Virology on the Net' hosts a page about bioscience jobs on the net. Its US-biased but could get you off the UK island 30/3/01*

TOOLS

Practice and practical

http://www.visions-of-science.co.uk *Science images competition, including entry forms and showing thumbnails of the winning images 1/11/00*

http://micro.magnet.fsu.edu/micro/gallery.html *Comprehensive science-based image library hosted by Florida State University 10/3/01*

http://www.culver.edu/homepages/Faculty/mjones/scipapr.htm *Elementary guidelines for writing scientific papers from Culver Stockton College, USA 26/3/01* http://micro.magnet.fsu.edu/primer/index.html *Comprehensive intro-duction to the use of the light microscope 5/3/00*

http://www.gen.latrobe.edu.au/microscopy/index.html *La Trobe University in Australia offers videos on the light microscope 18/3/01*

Software

http://www.passtheshareware.com *This site offers a range of software which can be used for improving the look of your screens 7/3/01*

http://www.pure-mac.com *This is a source of eleven Mac-based biology skills programs concerning molecular biology and ecology 7/3/01*

http://hotfiles.zdnet.com/cgi-bin/texis/swlib/hotfiles/search.html *Here are 57 free downloadable biology-relevant programs. This is a good-value site to visit for improving your skills in biology. From here I was able to go to a free download of software which showed how to cite bibliographic references 7/3/01*

http://www.shareware.com *Free downloads. Good starting point for various kinds of programs*

http://www.tucows.com *Sources of software, mainly to do with games and computing. From this site it was possible to get to a site at http://signup.domaindirect.com/cgi-bin/info.cgi?do=index&pageid=1028 where you can register a domain for $40 p.a. 7/3/01*

http://www.mirror.ac.uk *150 mirrors of sites of academic interest. This site mainly concerns software for computing – for example emailing and FTP downloads 7/3/01*

http://kissco.com *Cheap and cheerful downloads to monitor speed of your internet connection and tune them up 22/01/01*

http://www.softseek.com *Lots of free downloads for making your computer work better and faster 22/01/01*

http://www.qualest.co.jp/Download/KyPlot/kyplot_e.htm *Free download of a beta version of statistical software from Japan. Comprehensive but unproven. If you have commercial versions available, it might be best to stick to those (e.g. Minitab, SPSS, Statview) 21/01/01*

Software for the lab

http://www.totallab.com/html/download.htm *Software to buy for quantitatively analysing electrophoresis and other spots 1/12/00*

http://www.syngene.com/gt_prod_faq.asp *Apparently a bio-imaging company 1/12/00*

Citing internet sources

http://www.indiana.edu/~librcsd/eval/citing.html   *Advice on citation of sources of information 1/5/00*

General tools

http://www.hhmi.org/biointeractive   *Howard Hughes Medical Institute offers interactive pages concerning tools of modern biology, aimed at students 1/11/00*

http://www.wildflower.org.uk/guide/guide.htm   *National Wildflower Centre (UK) page for creating a wildflower meadow 10/3/01*

Dictionaries and glossaries

http://www.yourdictionary.com   *Excellent on-line dictionary and thesaurus which links to Merriam-Webster (see below) 5/3/00*

http://www.sciencekomm.at/advice/dict.html#life   *This site links to many on-line glossaries and dictionaries. It is a reasonable starting point but many of the links go to dictionaries that must be paid for 5/3/00*

http://m-w.com   *Merriam-Webster free on-line dictionary which includes scientific terms. There is also a thesaurus...but watch out for US spellings 11/3/01*

http://gened.emc.maricopa.edu/bio/bio181/BIOBK/BioBookglossC.html   *Glossary is supplied as part of an excellent, free on-line biology book 18/3/01*

Test your knowledge in biology

Objective question tests to gauge your understanding of biology.

http://www.discoverbiology.com/index/welcome.htm   *Objective question test related to a textbook which is partially on-line 30/3/01*

http://www.biology.arizona.edu/biochemistry/problem_sets/metabolism/metabolism.html   *University of Arizona test on metabolism (and other aspects of biology) 18/3/01*

http://www.cvxmck.edu.au/intranet/science/Biol/12es3.htm   *Clairvaux Mackillop College, Australia 30/3/01* http://resources.blackboard.com/scholar/sa/159/portal.jsp   *Free biology test questions are offered by the Blackboard Resource Center, a US commercial e-education venture. You can scale the size and nature of the test for yourself 22/7/01*

Miscellaneous searching

http://www.xe.net/cgi-bin/ucc/convert   *Currency conversion page 7/3/01*
http://search.yell.com/search/DoSearch   *Yellow pages for UK telephone numbers 7/3/01*
http://www.bt.com/phonenetuk   *Free directory enquiry of BT phone numbers 7.3.01*
http://www.fathom.com/index.jhtml   *Well-supported directory to on-line learning but courses have to be paid for 11/3/01*

SCIENCE FICTION

http://bailiwick.lib.uiowa.edu/earth2/index.html   *Mix of science and fiction based on a new US television series which has been called 'Wagon Train on another planet'. Only visit if you have the time to spare 11/03/01*

NEWS THAT MIGHT HAVE SOME SCIENCE IN IT

http://news.bbc.co.uk   *The keyword 'Biology' yielded 301 references. Not a pretty site but it is easy to navigate and there is lots of information 10/3/01*
http://www.cnn.com   *The keyword 'Biology' yielded 100 references, mostly with a particularly US slant 10/3/01*
http://www.telegraph.co.uk   *Cluttered web page of the Daily Telegraph. The keyword 'Biology' yielded 100 references 10/3/01*
http://www.ft.com   *Web page of the UK Financial Times. Rather surprisingly, the keyword 'ecology' in the global archive at http:// globalarchive.ft.com yielded over 120 references, most of which were relevant to an academic enquiry 10/3/01*
http://www.guardian.co.uk   *The Guardian web site which offered 75 references to 'Biology' ranging from links to a school-relevant web site*

on the human body at http://www.learning-connections.co.uk/curric/
cur_pri/h_body/frames.html to a link to the Human Genome Project
website at http://www.sanger.ac.uk/HGP 10/3/01

http://www.itn.co.uk *The UK ITN newservice web page. Unfortunately,
there were no matches in a search for 'Biology' at this site 10/3/01*

http://www.newsnow.co.uk *A news service which is updated every 5
minutes from different sources. 'Biology' yielded 11 references (US and
UK only). However, one of these told me that the Society of Actuaries
have concluded that testosterone 'plays havoc biologically and
behaviourally with men's bodies, leading to diseases and risk-taking
behaviour that are more common among men than women'. Indeed
10/3/01*

http://www.thetimes.co.uk *There is a good search engine at this site
hosted by The Times of London. It took me directly to a synopsis of the
molecular biology research which suggest that humans operate on the
basis of 30 000 functioning genes – at http://www.thetimes.co.uk/
article/0,,7-82727,00.html 10/3/01*

BOOKBUYING

http://www.bookbrain.co.uk *Blackwells-based web site which searches
web sites for books, giving full breakdown of costs and delivery charges
1/4/00*

http://www.amazon.co.uk *The original big on-line bookshop 1/4/00*

http://www.uk.bol.com *Books on-line 1/4/00*

http://www.alphabetstreet.infront.co.uk *On-line bookseller that sells
other stuff too 1/4/00*

http://s1.waterstones.co.uk/cgi-bin/wat01prd.storefront *Familiar high-
street name selling books on-line 1/4/00*

http://bookshop.blackwells.co.uk *Reputable scientific publisher book-
selling on-line 1/4/00*

MISCELLANEOUS

http://www.fensende.com/Users/swnymph *The Hunger Site. Click every
day. Sponsors pay for feeding the hungry in the developing world
30/3/01*

http://www.webpagesthatsuck.com *This site is all about web design.
Despite the title of the site, it contains useful information, for example
on creating a web site 1/4/01*

EXAMPLES OF INTELLECTUALLY DANGEROUS SITES

http://www.amazingdiscoveries.org/about.html   *This site purports to 'include studies on the credibility of the Bible, life after death, fulfillment of Biblical prophecies, current events, and other aspects of the gospel'. There is some solid biology here but given the mission statement, you need to be very careful using anything from this site. If you want to see the essence of the creationist argument, it is here 23/3/01*

http://www.onelife.com/evolve/index.html   *An electronics engineer writes about evolution. This site has a political perspective which colours the arguments. Handle with great care 1/4/01*

HOMEPAGE TOOLS

http://www.free-search.com   *Free icons and screensavers 3/6/00*

http://www.animfactory.com   *Free animations 3/6/00*

http://www.arttoday.com   *Free (and not so free) graphics 3/6/00*

http://www.clipart.com   *Free clipart, animations and graphics 3/6/00*

http://www.htmlgoodies.com   *Tutorials and banner exchanges etc. concerning web design 3/6/00*

http://www.netmag.co.uk   *Official site for .net magazine, with a lot of free downloads and contacts 3/6/00*

... AND FINALLY

http://home.att.net/~cecw/lastpage.htm   *The last page of the internet 5/3/00*

18
Web Sites for the School Syllabus

18.1   INTRODUCTION

The Association for Science Education in the UK has produced a useful brief guide to internet resources for science teachers (Sang, 2000). There is, however, a need for a more comprehensive review of web resources.

The 2000/2001 *Advanced Level school curriculum* of the Assessment and Qualifications Alliance (AQA) in England and Wales was used as a basis for producing a database of web-site URLs.

It was intended that a *student or teacher* who wished to find additional or support material would be able to go directly to the web site concerned, without having to use a search engine.

The search strategy involved using Google to obtain a list of sites which were then visited. These sites were included in the list according to whether the information:

- was at an *appropriate level*
- appeared to be *reliable*
- was produced *by staff* rather than students
- was *well produced* (no spelling errors, etc.)
- could be *generally understood* by an A-level student, though details might be more difficult to assimilate
- was *freely accessible* without subscription or other payment
- was presented in an *interesting* and visually stimulating way
- was on a web site that was likely to *persist* indefinitely.

Many web sites were encountered which did not satisfy these criteria.

18.2   KEYS FOR THE LIST

Each URL was classified as follows:

Key

Host – nature of the information provider
1   Private individual
2   School or university staff
3   Commercial
4   Governmental or inter-governmental organisation
5   Non-governmental organisation
6   Academic society or similar

Level – age of student in years
1   <11
2   11–16
3   16–18
4   18–21
5   >21

Links – are there links to other information external to the web site?   Yes/No

Contact – is the web page author directly contactable?   Yes/No

When transcribing URLs from this list to undertake a search, do not include text which is in italics.

[WEB SITES ON FOLLOWING PAGES]

Syllabus topic	Web URL	Where?	Host	Level	Links	Contact
General biology	http://gened.emc.maricopa.edu/bio/bio181/BIOBK/BioBookTOC.html	USA	2	2,3,4	Y	Y
Systematics						
Taxonomy	http://jrscience.wcp.muohio.edu/lab/TaxonomyLab.html	Miami, USA	2	3	Y	Y
Taxonomy and cladistics	http://www.cant.ac.uk/depts/acad/science/3rdyearcourses/ecol/ecol.htm	Canterbury, UK		3,4	Y	Y
Biodiversity	http://www.biodiv.org/doc/publications/guide.asp?lg=0	UN	4	4	Y	Y
	http://www.nhm.ac.uk/science/projects/worldmap/diversity/index.html	London	2	3,4	Y	Y
Diversity of micro-organisms	http://www.epa.gov/nerlcwww *(Environmental Protection Agency)*	EPA, USA	4	3,4	Y	Y
	http://www.microbelibrary.org/Curriculum/page2.htm	USA	6	3,4	Y	Y
	http://www.microbe.org *(American Society for Microbiology)*	USA	6	2,3	Y	Y
Life cycles	http://www.microbe.org/microbes/reproduction.asp	USA	7	2,3	Y	Y
	http://www.suite101.com/linkcategory.cfm/251/245	USA	2	3,4	Y	Y
Tools and methods						
Microscopy	http://www.mwrn.com/technical/camera.htm	USA	3	4	N	Y
	http://www.microscopy-uk.org.uk/mag/indexmag.html	UK	3	2,3,4	Y	Y
Chromatography	http://www.rpi.edu/dept/chem-eng/Biotech-Environ/CHROMO/chromintro.html	USA	2	2,3,4	Y	Y
Electrophoresis	http://www.mahidol.ac.th/mahidol/sc/scbc/electrophoresis.html	Thailand	2	4,5	N	N
	http://www.uni-giessen.de/~gh43/elphoexplain.html	Germany	2	2,3,4	N	N
Statistics						
Chi squared test	http://www.mste.uiuc.edu/patel/chisquare/intro.html	Illinois, USA	2	2,3,4	N	N
Standard deviation	http://www.bmj.com/collections/statsbk/2.shtml *(British Medical Association)*	BMA, UK	2	3,4	N	N

(continued)

Syllabus topic	Web URL	Where?	Host	Level	Links	Contact
Ecology						
Terrestrial	http://www.nrdc.org/land/forests/default.asp *(National Resources Defense Council)*	NY, USA	5	2,3,4	Y	Y
Aquatic	http://www.pbs.org/oceanrealm	USA	3	2,3,4	Y	N
Succession	http://library.thinkquest.org/17456/succession1.html	USA	6	1,2,3,4	Y	Y
Population biology	http://benthic.sc.edu/benthic/BIOL/Courses/BIOL301/Wethey/Outline09.html	USA	2	3,4	N	Y
Mark recapture 1	http://www.cnr.colostate.edu/~gwhite/mark/mark.htm	Colorado, USA	2	3,4	N	Y
Mark recapture 2	http://www.accessexcellence.org/AE/AEC/AEF/1995/nevin_grasshopper.html	Nat.Hlth.Mus., USA	4	2,3,4	Y	Y
Pyramids	http://www.marietta.edu/~biol/102/ecosystem.html#Pyramids5	Ohio, USA	2	3,4	Y	Y
Trophic levels	http://jan.ucc.nau.edu/~doetqp-p/courses/env470/Lectures/lec38/Lec38.htm	Arizona, USA	2	3,4	Y	Y
Cultivated plants	http://www.forages.css.orst.edu/Topics/Pastures/Species/Legumes/index.html	Oregon, USA	2	3,4,5,6	Y	Y
Cereals	http://millbake.crop.cri.nz/cyberguide/redir.htm?whgrain	New Zealand	4	3,4	Y	Y
Cycles	http://www.ae.iastate.edu/Ae573_ast475/Nutrient_Cycles_475.htm	Iowa, USA	2	3,4,5	Y	Y
Nitrogen cycle	http://www.notcatfish.com/chemistry/nitrogen_cycle.htm	Washington, USA	1	2,3,4	Y	Y
Pollution	http://www.worldwildlife.org/toxics *(WWF)*	Washington, USA	4	2,3,4	Y	Y
Conservation						
Management	http://www.naturenet.net	UK	5	1,2,3,4	Y	Y
Deforestation	http://www.ciesin.org/TG/LU/deforest.html *(Center for Earth Sci. Inf. Network)*	NY, USA	2	2,3,4	Y	Y
Grasslands	http://www.geribolsover.physiol.ucl.ac.uk/personal/steve/bov_home.html	UK	6	2,3,4	N	Y
Forest conservation	http://boto.ocean.washington.edu/eos *(Amazon Project, University of Washington)*	Washington, USA	2	2,3,4	Y	N

Topic	URL	Location				
Agriculture and environment	http://www.maff.gov.uk/erdp/docs/erdpdocsindex.htm (*DEFRA*)	UK	4	3,4	Y	N
Energy management	http://zebu.uoregon.edu/1998/ph162/14.html	Oregon, USA	2	3,4	Y	Y
Biochemistry						
Carbohydrates	http://www.harthosp.org/HealthInfo/scripts/scr0022.htm	Connecticut, USA	3	2,3,4	Y	N
Proteins	http://www.friedli.com/herbs/phytochem/proteins.html	USA	3	3,4	N	N
Lipids	http://gened.emc.maricopa.edu/bio/bio181/BIOBK/BioBookCHEM2.html#Organic molecules	Arizona, USA	2	2,3,4	Y	Y
Nucleic acids	http://esg-www.mit.edu:8001/esgbio/lm/nucleicacids/nucleicacids.html	Mass., USA	2	3,4	Y	Y
Enzymes	http://aci.mta.ca/Courses/Biology/BioWeb/2001catalysis.html#How	Sackville, Canada	2	3,4	Y	Y
Cell biology						
Introduction to cell biology	http://www.sheffcol.ac.uk/links/Science/Biology/Cell_Biology/	Sheffield, UK	2	3,4,	Y	Y
	http://www.learn.co.uk/guardianarchive/edulearn/education_resources/lesson_packs/science/lesson02/levels34.htm	London	3	1,2,3	Y	N
Cytology	http://www.cellsalive.com/index.htm	USA	3	2,3,4	Y	Y
Viruses	http://www.accessexcellence.org/AB/GG/examples_of_viruses.html (*Nat.Hlth.Mus.*)	USA	6	2,3,4	Y	Y
Prokaryotes	http://lsvl.la.asu.edu/mic494	Arizona, USA	2	1,2,3,4	Y	Y
Eukaryotes	http://www.ucmp.berkeley.edu/alllife/eukaryotamm.html	Berkeley, USA	2	2,3,4	Y	Y
Cell membranes	http://cellbio.utmb.edu/cellbio/membrane.htm	Texas, USA	2	2,3,4	Y	Y
Cell transport	http://esg-www.mit.edu:8001/esgbio/cb/membranes/transport.html	Connecticut, USA	2	2,3,4	Y	Y
Tissue structure and function						
Epithelium	http://www.usc.edu/hsc/dental/ghisto/epi/	California, USA	2	2,3,4	Y	Y
Blood	http://www.ultranet.com/~jkimball/BiologyPages/B/Blood.html	USA	3	2,3,4	Y	Y

Syllabus topic	Web URL	Where?	Host	Level	Links	Contact
Tissue structure and function (*continued*)						
Xylem	http://www.botany.hawaii.edu/faculty/webb/BOT410/Xylem/Xylem%2D1.htm	Hawaii, USA	2	2,3,4	Y	Y
Phloem	http://www.rrz.uni-hamburg.de/biologie/b_online/e06/06d.htm	Hamburg, Germany	2	2,3,4	Y	Y
Translocation in plants	http://botany.ru.ac.za/FACTFILE/F_INDEX.htm	Grahamstown, SA	2	2,3,4	N	Y
Organ structure and function						
Liver	http://www.usc.edu/hsc/dental/ghisto/gi/index-liver.html	USA	2	2,3,4	Y	Y
Blood vessels	http://gened.emc.maricopa.edu/Bio/BIO181/BIOBK/BioBookcircSYS.html	USA	2	2,3,4	Y	Y
Leaves	http://www.botany.hawaii.edu/faculty/webb/BishopWeb/BMW-7.htm	Hawaii, USA	2	2,3,4	Y	Y
Tissue fluid 1	http://atschool.eduweb.co.uk/middlecroft/tissue.htm	UK	2	2,3	Y	Y
Tissue fluid 2	http://www.fed.cuhk.edu.hk/~johnson/misconceptions/ce/misonceptions/mis_in_bio/circulation.htm	Hong Kong, China	2	2,3,4	Y	Y
Skeleton	http://www.eSkeletons.org	Texas, USA	2	2,3,4	N	Y
Muscles	http://www.gwc.maricopa.edu/class/bio201/muscle/mustut.htm	Arizona	2	2,3,4	Y	Y
Metabolism	http://ajpendo.physiology.org/cgi/content/abstract/280/2/E301	USA	6	3,4,5	N	N
Muscle contraction	http://www.accessexcellence.com/AB/GG/muscle_Contract.html	USA	3	2,3	N	N
Gas exchange	http://gened.emc.maricopa.edu/Bio/BIO181/BIOBK/BioBookRESPSYS.html	USA	2	2,3,4	Y	Y
Protists	http://www.ucmp.berkeley.edu/protista/basalprotists.html	USA	2	2,3,4	Y	Y
Gas exchange in fish and insects	http://gened.emc.maricopa.edu/Bio/BIO181/BIOBK/BioBookRESPSYS.html #Respiratory Surfaces	USA	2	2,3,4	Y	Y
Gas exchange in plants	http://www.ultranet.com/~jkimball/BiologyPages/G/GasExchange.html	USA	3	2,3,4	Y	Y
Blood gas transport	http://www.heartcenteronline.com/myheartdr/common/articles.cfm?Artid=375&startpage=2&#1	USA	3	2,3,4	Y	Y

Haemoglobin	http://www.ultranet.com/%7Ejkimball/BiologyPages/B/Blood.html#oxygen	USA	3	2,3,4	Y	Y
Digestion	http://www.nutristratcgy.com/digestion.htm	USA	6	2,3,4	N	N
Absorbtion	http://arbl.cvmbs.colostate.edu/hbooks/pathphys/digestion/smallgut/index.html	Colorado, USA	2	2,3,4	N	N
Respiration	http://esg-www.mit.edu:8001/esgbio/glycolysis/dir.html	MIT, USA	2	2,3,4	N	N
Respiration & photosynthesis	http://www.biosci.ohio-state.edu/~dcp/bio113a/ch910comp.html	Ohio, USA	2	2,3,4	N	Y
Photosynthesis 1	http://www.accessexcellence.com/AB/GG/photo_Resp.html	UK	3	2,3,4	Y	N
Photosynthesis 2	http://www.biology.arizona.edu/biochemistry/problem_sets/photosynthesis_1/01t.html	Arizona, USA	2	2,3,4	Y	Y
Heart	http://www.heartcenteronline.com/myheartdr/Articles_about_the_heart/The_Blood_Test_Center.html	USA	6	2,3,4	N	N
Heart function (virtual body)	http://www.ehc.com/vbody.asp	Nashville, USA	3	2,3,4	Y	Y
Homeostasis	http://calloso.med.mun.ca/%7Ethoekman/tempreg/tempreg.htm	Newfoundland	2	2,3,4	N	Y
Thermoregulation	http://www.ornithology.com/lectures/Metabolism.html	USA	1	2,3,4	Y	Y
Kidney	http://www.ultranet.com/%7Ejkimball/BiologyPages/K/Kidney.html	USA	3	2,3,4	Y	Y
Glucose	http://www.joslin.org/index.html	Boston, USA	2	2,3,4	N	N
Ammonia	http://www.notcatfish.com/chemistry/ammonia.htm	USA	3	2,3,4	Y	Y
Urea/uric acid	http://www.ultranet.com/%7Ejkimball/BiologyPages/U/UreaCycle.html	USA	3	2,3,4	Y	Y
De- and trans-amination	http://ull.chemistry.uakron.edu/Pathways/nitrogen_pool/index.html	USA	2	2,3,4	N	Y
Desert vertebrates	http://www.western.edu/faculty/jsowell/desertecology/ch4.html	Utah, USA	2	2,3,4	Y	Y
Desert insects	http://www.desertusa.com/mag98/july/papr/du_cicada.html	USA	3	1,2,3,4	Y	N
Xerophytes	http://learn.dccc.edu/~saquilan/ppt/leaf/sld020.htm	Delaware	2	2,3,4	Y	Y
Hydrophytes	http://botany.ru.ac.za/hydros.htm	South Africa	2	2,3,4	Y	Y
Transpiration	http://www.ianr.unl.edu/plantpath/peartree/homer/expos.skp/transpiration.htm	Nebraska, USA	3	2,3,5	Y	Y

(continued)

Syllabus topic	Web URL	Where?	Host	Level	Links	Contact
Food and diet						
Malnutrition	http://www.nutrition.org.uk/Facts/nutandhealth/malnutrition.html	UK	5	2,3,4	N	Y
Overnutrition	http://www.healthlibrary.com/reading/weight/index.htm	Bombay, India	5	2,3,4	Y	Y
Health	http://www.nlm.nih.gov/medlineplus/healthtopics.html *(Nat.Instit.Hlth.)*	USA	4	2,3,4	Y	Y
Additives	http://www.x-sitez.com/allergy/additives/colors100-181.htm	Australia	3	2,3,4	Y	N
Food storage	http://www.doh.gov.uk/busguide/hygrc.htm#summarytab *(UK Dept of Health)*	UK	4	2,3,4	N	N
Fermentation	http://www.bact.wisc.edu/microtextbook/Metabolism/Fermentation.html	Wisconsin, USA	2	3,4	N	Y
Brewing	http://www.alabev.com/history.htm	Colorado, USA	6	2,3,4	N	Y
Baking	http://enzymes.novo.dk/cgi-bin/bvisapi.dll/discover/discover.jsp	Denmark	6	2,3,4	N	Y
Ruminants	http://www.gov.nf.ca/agric/pubfact/livestoc/LSD037.HTM	Newfoundland	4	2,3,4	Y	N
Growth and development						
Plant hormones	http://www.luc.edu/depts/biology/dev.htm	Chicago, USA	2	2,3,4	Y	Y
Plant growth experiments	http://tidepool.st.usm.edu/crswr/111planthormones.html#1	Mississippi, USA	2	2,3,4	N	Y
	http://www-saps.plantsci.cam.ac.uk/sapshom.html *(School level)*	Cambridge,UK	2	1,2,3	Y	Y
Plant growth	http://www.dpw.wageningen-ur.nl/pdmm/2-plant/peconcpt/pec000fs.htm	Netherlands	2	2,3,4	N	Y
Mineral nutrition in plants	http://149.152.32.5/Plant_Physiology/minerals.html	Connecticut, USA	2	2,3,4	N	Y
Insect development	http://dnr.state.il.us/lands/education/classrm/wingleg/DEVELOPM.HTM	Illinois, USA	2	2,3,4	N	Y
Insect meta-morphosis	http://www.devbio.com/chap18/link1803a.shtml	Pennsylvania, USA	2	2,3,4	N	Y
Coordination						
Gut hormones 1	http://human.physiol.arizona.edu/SCHED/GI/Lynch69/Lynch.169.html	Arizona, USA	2	3,4	N	Y
Gut hormones 2	http://soma.npa.uiuc.edu/labs/greenough/statements/rswain/hormones/012496.html	Illinois, USA	2	2,3,4	N	Y

Topic	URL	Location				
Nervous coordination	http://www.ultranet.com/~jkimball/BiologyPages/S/Synapses.html	USA	3	2,3,4	Y	Y
Sensory physiology	http://www.medfak.uu.se/fysiologi/Lectures/GenSensPhys.html	Uppsala, Sweden	2	3,4	N	Y
Pacinian corpuscle	http://www.medfak.uu.se/fysiologi/Lectures/Pacini.html	Uppsala, Sweden	2	2,3,4	N	Y
Autonomic nervous system	http://faculty.washington.edu/chudler/auto.html	Washington, USA	2	2,3	Y	Y
Eye	http://www.exploratorium.edu/learning_studio/cow_eye (*Dissection offered by a museum*)	San Francisco, USA	2	2,3	Y	Y
Retina and eye structures	http://retina.anatomy.upenn.edu/~lance/eye/eye.html	Pennsylvania, USA	2	2,3,4	N	Y
Retinal focusing	http://www.nyee.edu/page_deliv.cgi?Page_ID=8&linker=0&faq_id=26 (Hospital)	NY, USA	2	2,3,4	N	Y
Ear	http://ctl.augie.edu/perry/ear/ear.htm	Sioux Falls, USA	2	2,3,4	N	Y
Behaviour (in farm animals)	http://www.liru.asft.ttu.edu/EFAB/default.asp	USA	4	2,3	N	Y
Conditioning	http://users.csbsju.edu/~tcreed/pb/pavcon.html	Ohio, USA	2	2,3,4	N	Y
Thyroxine	http://www.waichung.demon.co.uk/webanim/T4Web.htm	Glasgow, UK	2	2,3,4	N	Y
Thyroid	http://www.the-thyroid-society.org/faq/	USA	3	2,3,4	N	N
HRT	http://www.bidmc.harvard.edu/obgyn/menopause_hormone.asp#topics	Boston, USA	2	2,3,4	N	N
Classical genetics	http://www.bact.wisc.edu/MicrotextBook/BactGenetics/geneticterms.html	Wisconsin, USA	2	3,4	N	Y
History of genetics	http://129.128.91.75/de/genetics/70gen-history.html	Alberta, Canada	2	2,3,4	N	Y
Cell cycle	http://gened.emc.maricopa.edu/bio/bio181/BIOBK/BioBookmito.html	USA	2	2,3,4	Y	Y
Mitosis	http://www.biology.arizona.edu/cell_bio/tutorials/cell_cycle/cells3.html	Arizona, USA	2	2,3,4	Y	Y
Meiosis	http://www.biology.arizona.edu/cell_bio/tutorials/meiosis/main.html	Arizona, USA	2	2,3,4	Y	Y
Conception	http://www.healthsci.utas.edu.au/weller/+docs/repro6.htm#Results	Tasmania, Australia	2	2,3,4	N	Y

(continued)

Syllabus topic	Web URL	Where?	Host	Level	Links	Contact
Classical genetics						
Fertilisation	http://www.doctorsaab.com/asppages/marriage2-sex.asp	India	5	2,3,4	Y	Y
Mutations	http://www.bact.wisc.edu/MicrotextBook/BactGenetics/mutFreq.html	Wisconsin, USA	2	3,4	N	Y
Mendelian inheritance in man	http://www3.ncbi.nlm.nih.gov/htbin-post/Omim/getmim	NIH, USA	4	3,4,5	N	N
Autosomal linkage	http://www.uic.edu/classes/bms/bms655/lesson12.html	Illinois, USA	2	2,3,4	N	Y
Recombination	http://www.accessexcellence.com/AB/BC/Genetic_Recombination.html	USA	3	2,3,4	Y	N
Hardy–Weinburg	http://library.thinkquest.org/18757/further.htm (*Produced by students, but is good*)	South Australia	2	2,3,4	N	Y
Artificial selection	http://faculty.uca.edu/ben.waggoner/biol4415/lect4a/sld001.htm	Arkansas, USA	2	2,3,4	N	Y
Molecular biology						
Genetic code	http://vector.cshl.org/dnaftb (Cold Spring Harbor)	NY, USA	2	2,3,4	Y	N
DNA	http://vector.cshl.org/dnaftb/15/concept/	NY, USA	2	2,3,4	Y	N
RNA	http://www.iacr.bbsrc.ac.uk/notebook/courses/guide/rnast.htm#Com	UK	4	2,3,4,5	Y	Y
Replication	http://esg-www.mit.edu:8001/esgbio/dogma/repl.html	Mass, USA	2	3,4	Y	Y
Protein synthesis	http://www.visionlearning.com/library/science/biology-1/BIO1.1-nucleic_acids.htm	USA	1	2,3,4	N	Y
PCR	http://www-biology.ucsd.edu/others/dsmith/classes/pcr.html	San Diego, USA	2	2,3,4	N	N
Enzyme biotechnology 1	http://www.fst.rdg.ac.uk/courses/fs560/topic2/t2index.htm	Reading, UK	2	2,3	Y	Y
Enzyme biotechnology 2	http://www.enzymes.co.uk/immobilisedenz.htm	Reading, UK	2	3,4	Y	Y
Lac operon	http://www.people.virginia.edu/~rjh9u/lacoperonanim.html (*Animation*)	Virginia, USA	2	2,3,4	N	
Human genome project	http://www.sanger.ac.uk/HGP/overview.shtml	Cambridge, UK	4	3,4,5	N	N
Gene technology	http://www.wnet.org/innovation/show1/html/animation2.html	Virginia, USA	3	2,3,4,5	Y	
DNA fingerprinting	http://www.biology.washington.edu/fingerprint/whatis.html (*By students; well presented*)	Washington, USA	2	2,3,4	N	N

Recombinant DNA technology	http://www.accessexcellence.org/AB/BA/casestudy3.html (*Accessexcellence*)	USA	3	2,3,4	Y	N
Cloning	http://www.ri.bbsrc.ac.uk/library/research/cloning/glossary.html	Edinburgh, UK	3	2,3,4		
Reproduction						
Asexual reproduction in 5 kingdoms						
I Monera ('Bacteria')	http://www.mtsu.edu/~sharlow/cellcycle_web/sld006.htm	Tennessee, USA	2	2,3,4	N	Y
II Protista ('Protozoa')	http://www.tulane.edu/~wiser/protozoology/notes/INTRO.html#top	Tulane, USA	2	2,3,4	N	Y
III Fungi	http://perth.uwlax.edu/biology/volk/fungi3/sld045.htm	Wisconsin, USA	2	2,3,4	Y	Y
IV Lower plants	http://www.sbs.auckland.ac.nz/biology_web_pages/nzplants/moss_asexual_reproduction.htm	Auckland, NZ	2	2,3,4	N	Y
IV Higher plants	http://web.ukonline.co.uk/webwise/spinneret/pot/asex.htm	UK	1	2,3	N	Y
V Animals	http://www.britannica.com/eb/article?eu=118973&tocid=25731	USA	3	2,3,4	Y	N
Sexual repr. in flowering plants	http://www.sbs.auckland.ac.nz/biology_web_pages/nzplants/angiosperms_sexual_reproduction.htm	Auckland, NZ	2	2,3,4	N	Y
Sexual repr. in humans	http://www.ucalgary.ca/UofC/eduweb/virtualembryo/humans.html	Calgary, Canada	2	2,3,4	Y	Y
Birth	http://pregnancy.about.com/health/pregnancy/library/weekly/aa033098.htm	USA	3	2,3	N	N
Caesarean birth	http://www.fensende.com/Users/swnymph/csect/gallery.html	USA	1	2,3,4	Y	Y
Pregnancy	http://www.epregnancy.com/directory/	USA	3	2,3,4,5	Y	Y
Lactation	http://mammary.nih.gov (*Laboratory of Genetics & Physiology, National Institute of Health*)	Maryland, USA	2	2,3,4	N	Y

(continued)

Syllabus topic	Web URL	Where?	Host	Level	Links	Contact
Evolution						
Natural selection	http://www.athro.com/evo/evframe.html	Penn., USA	3	3,4	N	Y
	http://www.homeworkhelp.com/homeworkhelp/freemember/text/bio/high/private/ch08/0303/main.htm	California,USA	3	2,3,4	N	N
Isolation	http://arnica.csustan.edu/biol1010/evolution/evolution.htm	California, USA	2	3,4	N	Y
Speciation	http://arnica.csustan.edu/biol1010/speciation/speciaton.htm	California, USA	2	3,4	N	Y
Antibiotic resistance	http://whyfiles.org/038badbugs/mechanism.html	Wisconsin, USA	2	2,3,4	N	N
Pesticide resistance	http://ipcm.wisc.edu/pubs/pest/a3615.htm	Wisconsin, USA	2	2,3,4	N	N
Microbiology						
Culture techniques	http://www.engr.umd.edu/~nsw/ench485/lab8b.htm#Introduction	Maryland, USA	2	3,4	N	Y
Patterns	http://www.slic2.wsu.edu:82/hurlbert/micro101/pages/101hmpg.html	Washington, USA	2	2,3,4	Y	Y
Fermentation	http://biology.clc.uc.edu/courses/bio104/cellresp.htm	Cincinnati, USA	2	3,4	N	Y
Fermenters	http://www.bbc.co.uk/science/tw/items/990303_highpressurefood.shtml	UK	4	2,3,4	Y	Y
Bioreactors	http://www.cfe.cornell.edu/compost/garbagecans.html	NY State, USA	2	2,3,4	Y	Y
Growth of cultures	http://www-micro.msb.le.ac.uk/LabWork/bact/bact2.htm	Leics., UK	2	2,3,4	Y	Y
Biotechnology						
Isolated enzymes	http://www.biores-irl.ie/biozone/index.html	Ireland	3	2,3,4	N	N
	http://enzymes.novo.dk	Denmark	6	2,3,4	N	Y
Biogas	http://www.roseworthy.adelaide.edu.au/~pharris/biogas/beginners	Adelaide, Australia	2	2,3,4	Y	Y
Gasohol	http://www.cleantechindia.com/neweic/Gasohol.htm	India	4	2,3,4	N	N
Waste treatment	http://kola.dcu.ie/~enfo/bs/bs28.htm	Dublin, Ireland	3	2,3,4	N	Y
Pathogenicity	http://www.ntu.ac.uk/life/sh/modules/hlf349/349.htm	Wisconsin, USA	2	3,4	Y	Y
Smoking-related disease	http://www.nlm.nih.gov/medlineplus/smoking.html (*US National Library of Medicine*)	USA	4	2,3,4	Y	N

Topic	URL	Location				
Plant pathogens	http://www.outreach.uiuc.edu/PPP/welcome.htm	Illinois, USA	2	2,3,4	N	Y
Microbial pathogens	http://www.bact.wisc.edu/Microtextbook/disease/introduction.html	Wisconsin, USA	2	3,4	Y	Y
Parasites	http://ucdnema.ucdavis.edu/imagemap/nemmap/ent156html/vertcom	California, USA	2	3,4,5	Y	Y
Natural defence mechanisms						
Immunity	http://www.biology.arizona.edu/immunology/tutorials/antibody/structure.html	Arizona, USA	2	2,3,4	Y	Y
	http://gened.emc.maricopa.edu/Bio/BIO181/BIOBK/BioBookIMMUN.html#The Lymphatic System	USA	2	2,3,4	Y	Y
Neutrophils in pus	http://www.pathguy.com/histo/038.htm	Kansas, USA	2	2,3,4	N	Y
Allergies	http://allergies.about.com/health/allergies/mbody.htm	USA	3	2,3,4	Y	N
Antibiotics	http://helios.bto.ed.ac.uk/bto/microbes/penicill.htm	Edinburgh, UK	2	2,3,4	N	N
Insulin	http://www.diabetes.ca/about_diabetes/insulin.html	Canada	5	2,3,4	Y	Y
Bioengineered insulin	http://www.drinet.org/html/genetic_engineering.htm	USA	5	2,3,4	Y	Y
Antibodies	http://www.accessexcellence.com/AB/GG/antiBD_mol.html	UK	3	3,4	N	N
Monoclonal antibodies	http://www.ultranet.com/~jkimball/BiologyPages/M/Monoclonals.html	USA	3	2,3,4	Y	Y
Use of microorganisms in medicine	http://www.md.huji.ac.il/microbiology/book/ch007.htm	Jerusalem, Israel	2	3,4,5	N	Y

19
Conclusion

The internet is an extension of human knowledge. Since human knowledge is increasing exponentially, the internet is also increasing exponentially. And the process is *synergistic*. The existence of the internet allows humans to acquire new information at a faster rate than if there were no internet. For example, home computers spend a lot of time doing nothing. The power of some of these idling computers is being harnessed via the internet to find out if there are any other forms of sentient life in the universe. With the agreement of their owners, astronomical information is fed to these computers over the internet. While their owners are at work or asleep, the computers analyse the data to look for signs of life.

The net is becoming vital as an *educational tool*. Resources in libraries, universities and industry are becoming accessible in new ways and to a new audience. Prof. Bruce Royan has suggested that the role of the teacher will change from being a 'sage on the stage' to being a 'guide to the sites'. In the past, the best teachers have always been *learning* facilitators rather than founts of knowledge and this situation may devolve to all teachers in the future.

As wireless applications develop, the information will become increasingly immediate and accessible. The dangers of overload will become greater and users will have to develop a greater acuity. They will need to be able to *mine* the knowledge mountain, to *scan* it and *filter out* the extraneous information, focusing on the high-quality information which is most relevant to their needs.

As more information is devolved to the internet, its strengths become its weaknesses. Users have access to both works of value and valueless work alike, and the use is left to their own devices in trying to distinguish worth. 'Nobody plays God on the internet' (Naughton, 1999). The challenge to distinguish worth has always been true of the relationship between people and information but the immediacy of access offered by the internet is a new issue. As with any technology, users must know what they are dealing with, must be wary of it, must become skilled in its use and must be circumspect about what it can do for them. Naughton also says, 'The truth is that the net is wonderful in what it can do for us and terrifying in what it might do to us'.

However, there is another side to the electronic information explosion. Some of the initial ethos of openness on the web is being lost. University sites are closing down their web pages so that password access is necessary. For example, for some years, a UK university had a web site which gave excellent advice on citation of web pages, open to all. Now, non-members of the university are locked out. Journals are demanding subscriptions for complete articles and education resource providers expect payment. There are often loss-leader pages on the net, where a resource is freely available for some months but then locks out those who do not accept newly imposed charges.

It is gratifying that some individuals and institutions are willing to make their intellectual property freely available to all. Naughton convinces us that '*Openness is the net's greatest strength and the source of its power*'. There are certainly both general and specific benefits to such openness. Generally, everyone benefits from the fact that the information is so readily available. Specifically, such institutions will receive many virtual visitors, their reputation will be enhanced and they will increase their business as a result. Some, particularly USA-based, universities have been quick to understand this principle and are reaping benefits as a result.

Appendix

Getting the Computer Going

HARDWARE BASICS

The basic computer system (Fig. A1) comprises a:

- *computer* containing
    - ⇒ processor running at approximately 400 MHz
    - ⇒ memory for running programs – approximately 64 Mb of RAM (random access memory)
    - ⇒ hard disc drive for storing programs: 6 gigabytes
    - ⇒ CD-ROM drive usually 24-speed (or DVD drive)
    - ⇒ sound card which is Soundblaster-compatible 128
    - ⇒ graphics card with 8 Mb of memory for using images
    - ⇒ floppy disc drive for carrying information and programs between computers
    - ⇒ modem operating at 56 kilobytes per second (kbps)
    - ⇒ connecting ports
- *monitor* – the screen which is sometimes called a *visual display unit* (*VDU*)
- *keyboard*
- *mouse*

Usually the connections are at the back of the main case (Fig. A2). In a laptop or notebook computer, all of these are combined in a single case.

To be connected to other computers, the modem will have to be connected to a *telephone line*.

Figure 1   The basic hardware

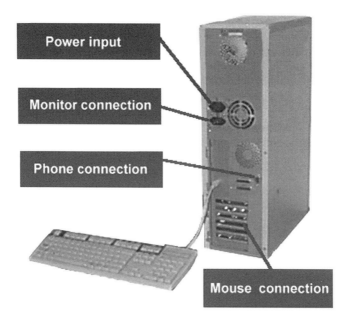

Figure 2   The connections at the back of the main case

The monitor allows you to see what the computer is doing and also allows you to give the computer instructions. Most programs now are icon-driven via the *graphic user interface* (GUI) which is a system pioneered by Apple Macintosh computers. In other words, there are a series of small pictures (*icons*) on the screen and you move a *cursor* (a kind of locality indicator which focuses your attention at any location on the screen) to sit on one of the small pictures. When the cursor sits on the icon, you give a start command and whichever program is linked to the icon will start to work ('*run*'). The cursor is moved by one of the following:

- mouse
- joystick
- roller ball
- stroke mat

Each of these methods has its own advantages and each requires different kinds of physical skills. With overuse, each will cause you a different kind of repetitive strain injury in the various joints of your arm and hand. It is becoming increasingly possible to give word-of-mouth instructions to the computer; this will undoubtedly get much more important in the future as computers get more intelligent.

The mouse is currently the most common method of moving the cursor. On the mouse are two buttons. Most of the start commands are given by a *rapid* double-click with the index finger on the *left mouse button*.

Frequently, more up-to-date systems have useful and interesting commands connected to the right mouse button. For example, if you have spelt a word wrongly, and the computer is set up appropriately, the mis-spelled work is indicated by a red underline. Place the cursor on this word, click with the *right mouse button* and a set of alternative correct spellings will be shown. Move the cursor to the one you want, click with the left mouse button and the word will be corrected automatically. This can save an enormous amount of time, despite the fact that some of the spellings in the computer's dictionary are wrong.

You may also find it useful to have a *printer* and a *scanner*. These types of equipment are not strictly necessary but are usually available. Any equipment which is extra to the main functioning of the computer has the general name of '*peripherals*'.

SOFTWARE BASICS

The computer runs using *programs* which are written in *machine code*. This is difficult for humans to interpret, so there are programming languages which allow the machine to be told what to do more easily. These include languages such as Basic, C+ and Pascal. Even so, it can be inconvenient to direct the computer through these languages, so a user-friendly platform is built, using these languages. Typical of these are Microsoft Windows, Linux and the Apple systems. It is probably fair to say that the most common is Windows, though it uses GUI ideas originally pioneered by Apple. Once you are using the platform, you can then use programs which are called 'applications' – for example, word processing programs such as Word, data handling programs such as Excel, database programs such as Microsoft Access and internet search programs such as Netscape. These programs have the general title of 'application software'.

Increasingly, computers are starting to learn from the user and try to automatically correct aspects such as the layout of documents. Frequently, the computer gets it wrong, which can be irritating and frustrating. To wrestle control away from Word, for example, use the toolbar at the top of the screen. Go to *Tools>Autoformat>Autocorrect as you type*, and un-tick all the boxes.

SPECIAL NEEDS

There are a variety of adaptive and alternative options, including *voice recognition* and *Braille* input and output systems. These are reviewed in Lynas (2000).

Glossary

address – unique location of an internet web page

ADSL – asynchronous digital subscriber line – a fast method of transferring information, via a modem, to and from the internet. Current operation speed is 1024 kb/s

antivirus program – protects your computer against viruses, but needs to be constantly renewed to keep up with new viruses

application – computer program which actually does something. An example is the word-processing package Word. Applications are seen as being different from programs such as Windows which represent an environment in which applications can operate

attachment – file that is appended to an email message

autostart – short-cut programs which run in the background of your computer. In Windows, they are displayed at the bottom right of the screen

beta – software which is in its final but as yet uncompleted stage of development. It may contain bugs

bit – the basic unit of computer data

      one byte = 8 bits

      kb = 1000 bytes

      Mb = 1 000 000 bytes

      Gb = 1 000 000 000 bytes

bookmark – Netscape's word for preserving a link to a particular web page that you might want to visit again

broken link – a link which is not working properly

browser – a program which searches the internet for you. The most common ones are Internet Explorer and Netscape

bug – an error in a program

cache – temporary memory into which recently visited web pages are downloaded and saved

caveat – a friendly warning (Latin)

chat room – a monitored web site where a group of people post messages and thus hold a discussion

citation – a reference to information which is, or has been, available elsewhere. Citation usually means an author and date reference in the text, with a full reference at the end of the work. In some formats, it means a superscripted number in the text and then the full reference as an endnote

client – the computer which requests information. When it supplies information, it is a server

clip art – a freely available collection of images which you can copy into your documents

cogent – relevant, useful, just what you are looking for

cognoscenti – people on the inside of a specialised clique

compression programs – used to wrap up a program so that it takes as little memory as possible. Such programs need to be unwrapped before they can be used

configuration – all the settings in your computer that allow it to work. When you become more skilled, you can configure the machine to suit your requirements

copyright – the system which protects the authors or other owners of a piece of intellectual property against plagiarism

cookies – information about how you have used the web. This information is stored on your computer and is observed by a web site with which you have been in contact

cursor – the arrow, bar or other icon which indicates the focus of your attention on the screen. It is usually controlled via a mouse

dead links – hypertext links which no longer work. This usually happens because the target web page has been deleted or moved. Dead links usually yield a 404 error message. If it is important enough, try to re-find the link by dissecting the URL

decryption – the decoding of information which was encrypted to keep it secret and secure

default – a fall-back position. For example, your computer is set up for a default browser. You might use other browsers but the computer will eventually switch back to the default

dialog box – usually a drop-down window which appears when you click on an icon. It offers you options in the form of a sub-menu

directory – there are different kinds of directories. The file manager of your computer will allow you to store files in directories. A web directory is a list of web sites. Directory can mean the same thing as folder

dissection of the URL – serially remove domains from the right-hand side of a URL and hit 'return' until you reach something useful

domain name – host name followed by top-level category (e.g. .org, .co, .com) and then a series of other categories, the first being a country category (e.g. .UK). Every net address is part of a domain net system which comprises a series of hidden digits

download – method of getting files from the web onto your computer

DSL – digital subscriber line transmission; currently the most common method of connecting to the internet

email – a messaging service that allows you to exchange messages via phone lines and other communication links. Email means electronic mail

emoticons – (smileys) – punctuation which expresses emotions. They used to be assembled piecemeal, e.g.
: –) (look at it sideways) . Your computer now be set up to automatically replace :-) with ☺

encryption – a way of coding data so that it can't be read by an outsider; it is a protection against fraud

esoteric – describes information which is only understood by the specialised few, by the *cognoscenti*

FAQ – frequently asked questions are a method of helping you to solve problems without having to contact the supplier directly or personally

Favorite – this is the Microsoft equivalent of the Netscape Bookmark. It is a way of retaining the address of a web page that you might want to visit again

files – blocks of computer meory which contain programs or data in one form or another

filters – programs that allow you to remove unwanted emails, usually spam, before they get to you

firewall – security protocols developed for institutions such as universities that screen incoming information and try to keep out problems such as viruses

flash – a plug-in which allows you to display animations

folders – bits of memory in which files are stored. Has the same meaning as 'directories'

form – on some email pages, you are asked to send information to the web site. This is done by filling in boxes on the form

frame – this is a part of a web page. Imagine that the web page was constructed as a set of tables. Each table would be a frame

freeware – programs which are available free of charge

FTP – a file transfer protocol is a fast method of uploading and down-loading files

gateway – a computer that controls access to the internet

generic – describes something which is common to many circumstance, something which is general or global gif (graphic interchange format) – this is a graphic image format which is economic on memory and can be viewed by a browser

hard disc – this is the software storage location for your computer. It is a disc, semi-permanently installed inside the computer. Capacity is measured in gigabytes (Gb) or megabytes (Mb)

hit – this the URL which results from a search. If your search yields 12 URLs, that is 12 hits

homepage – this is the major page of a web site, or it can be your own individual page (in other words the primary page of your own web site if you had one)

html – the code that tells your browser how to decode and display a web page. It stands for 'hypertext mark up language'

http (hypertext transfer protocol) – the program which carries information between your computer and the internet. For modern browsers you don't need to use http:// but can start with the www. part

hyperlink – this is a way of jumping from one web page to another by clicking on an appropriately highlighted word or image. The highlighted word is usually underlined and often changes colour after you have clicked on it

icon – an image which is usually hyperlinked to something

ICQ – I seek you. A way of finding people on the internet

image file – a file which is made up of an image. The most primitive, memory-consuming image is in the form of a bitmap (.bmp). More economical image files are .jpg and .gif files

IMAP – internet message access protocol. A program for receiving and passing on email messages

installation – conversion of downloaded software from the internet so that it will work on your computer

instant messages – emails which go to one person rather than several

interface – the junction between the computer and you. It is the menus, highlights and boxes on which you click for things to happen

internet – this is the total of the linked computers which contain hardware and software which allow humans to talk to each other electronically. It resembles a web of connections and is sometimes called 'the web'

internet relay chat channels (IRC) – these allow you to hold live conversations with a chat group. Examples are WSIRC for PCs and IRCLE for Macs

internet service provider (ISP) – these are businesses which operate computers and software that allow you to communicate with the internet. Your modem dials a special access number for the modem of your ISP. If the ISP number is on-line and not busy, the connection between your computer and the internet is made

intranet – networked computers in a particular locality or institution or building

IRC – internet relay chat is a method of holding a live conversation with others over the internet

ISDN – integrated services digital network uses digital rather than analogue technology to connect computers and can therefore handle large amounts of information very quickly

iterative, iteration – this is the process of cyclically doing something. Cleaning a window in a house is an iterative process of wiping, checking and wiping until the window is clean

Java – a web-based programming language similar to, but more powerful than, html. Because it is more complex, it can cause problems with older browsers

Javascript – a programming language which lies within Java. It allows you to do things like change the colour of an icon when the cursor moves over it

jpeg (joint photographic experts group) – this is a graphic image format which is particularly useful on the internet because it is economic on memory and is downloaded quickly

keyword – the word or phrase which is entered into a search box in a search engine

link – a bridge between separate web pages; it is a type of 'hot spot' which allows you to go further

location bar – the long white bar at the top of a web page which tells you which web page you are on. If you want to go to a new web page, the web address is typed here

login – this is a way of getting access to a program and usually involves you entering a username and password that you have been given

mailbox – your personal place on the ISP server where your emails are stored

mailing list – a list of email addresses; when you send an email to a mailing list, the message goes to everyone on the list

mail server – the computer and software used by your ISP to handle your emails

menu – a list of hyperlinked options. Choose one and click and you will be taken to the next destination

modem – the box of tricks which connects your computer to the internet via a phone line. It needs to be fast enough. A minimum specification is V.90, which receives at 56 kilobytes per second

monitor – the 'television screen' on which your web pages and computing actions are displayed

mouse – the device which allows you to move the cursor across the screen

navigation – this is the process of using hyperlinks to move around a web site, or from web site to web site across the internet

netiquette – good, polite behaviour on the internet

newsgroup – this is a subject-specific place on a server to which people can send emails so that they can be read by others and maintain a discussion

off-line – you are working on your computer but it is not connected to the internet. It saves money if you can construct emails off-line, then connect to the net (go on-line) to send them. The sending is almost instantaneous

on-line – your computer is connected to the internet and you may be paying for the minute-by-minute cost of the connection

on-line services – ISP that supplies both access to the internet and a whole range of other electronic communication facilities – for example America On line (AOL), Compuserve, Microsoft Network (MSN), LineOne

optical sensor – the equivalent of the roller ball in a mouse

Outlook Express – a Windows-based email program which is part of Internet Explorer

parasitoid – something between a parasite and a predator. Usually it is an insect predator which is smaller than its prey, for example a solitary wasp

parsimony, parsimonious – economical, efficient. Literally, best value for money

pdf files (portable document format) – these are files which store scanned copies of text and images in an economical way. They are one of the preferred ways of storing institutional documents such as government reports. They are created and read by using programs such as Adobe Acrobat. It is easy to download readers for free, but you have to pay for pdf writers. Get an Adobe Acrobat reader free from http://www.adobe.com/acrobat

peer review – the system which guarantees the quality of published science. Any article which is to be published in a peer-reviewed journal is usually sent to three independent academics who will (separately and usually anonymously) produce hyper-critical reviews, recommending or denying publication

plagiarism – the theft of language, wording, images or ideas; it is a serious academic misdemeanour. Plagiarism is an intellectual crime

plug-in – this is an extra piece of software which you usually download. It enhances the things your computer can do with web information. A typical free plug-in is Quicktime (http://www.apple.com/quicktime) which allows you to see video images on-line

POP3 (post office protocol 3) – program used by your email program to download emails from your ISP

pop-up – a piece of information, usually an advertisement, which appears on the screen of its own accord

port – a socket at the back of the computer through which electronic information will pass. There are several such sockets

portals – a collection of links which refer to a specific subject. Portals often supply a range of services including web access, chat rooms and email

postmaster – person at an ISP who manages the email system

précis – to summarise something in your own words; this is best done by reading a paragraph, covering it over and then writing a summary of what you remember from the paragraph

processor – the part of your computer that does the 'thinking'. The speed of its activity is measured in megahertz (MHz); more than 350 MHz is good

program – the coded instructions used by the computer which allow something to happen

provenance – where something comes from. Your provenance is your parents and the place you were born. The provenance of a web page is the person who produced it and the place it originally resided

PS/2 port – this is a connection socket on the back of the computer. These are circular. One is for the keyboard and the other is for the mouse. Make sure these are plugged in before you switch on the computer

public domain – material which anyone who has access to the internet can look at

RAM (random access memory) – the memory which is used by the processor and the hard drive to run all the different programs. Data are moved in and out of the RAM onto and off the hard drive under the control of the processor; 64 megabytes (Mb) is good

real time – the speed of real life. An email sent now may bounce around the world for some time. It won't arrive in real time. If it goes immediately to its destination, it happens in real time

rectangularity – the way frames on a web page are arranged in the form of a hidden table, made from edge-to-edge rectangles

registry – settings in your computer. In general, leave these alone if you get access to them

resolution – the sharpness of an image, dependent on the number of pixels. The more pixels, the bigger the picture, the sharper the resolution and the more memory that is taken up

scrollbar – a long thin icon at the side or bottom of the screen with a slider button and up and down arrows. You can move round the screen by:

- clicking and sliding the slider button up and down
- move one line at a time by clicking on the up and down arrows
- clicking on the scrollbar itself; large parts of the page will be traversed

search engine – a web site that is engineered to look for other sites on the internet using keywords

secure server – a server which has special encryption facilities to prevent messages being illicitly read

serendipity, serendipitously – finding something by chance. This is one of the pleasures of the internet

serial port – an old-fashioned port, used to connect to the mouse on some older computers. It has nine pin holes

server – large, powerful computer operated by an internet service provider. Subscribers are allowed to store their home pages on these servers. There are web servers, email servers and news servers. The server computer passes information *to* the internet. The client computer takes information *from* the internet

shareware – programs that are available free of charge. Often they are only available for a limited period, then you have to pay

signature – text you put at the bottom of an email which tells the reader who you are

smileys – emotion messages (emoticons) such as these sideways faces :-) and :-(

SMTP (simple mail transfer protocol) – elementary programs used by the servers to exchange emails

spam – electronic version of unwanted junk mail, often along the lines of 'Claim your prize and get rich quick'

SSL (secure socket layer) – encryption that recognises surfers via encryption certificates

surfers – people who are using the internet; more specifically, people who are casually visiting web sites and following links which seem of interest

swap files – temporary, virtual memory files used by the computer when it starts to run out of memory

synergism, synergistic – several things have a multiplicative effect on each other. For example, one pollution makes another pollution worse than if it were operating alone; or a diligent friend will make you diligent, and then you make them even more diligent

synopsis, pl. synopses – summary produced at a particular point in time. A synoptic view is a snapshot view

temporary internet files – Internet Explorer stores data here temporarily. It is the equivalent of a cache

thumbnails – small images on a web page that are hyperlinked, frequently linked to larger versions of the same image. A double click on the thumbnail will take you somewhere

touchpad – equivalent to a mouse found on notebook portable computers. The pad is stroked to move the cursor, and tapped to give the equivalent of a mouse button click

upload – transferring data from your computer to another computer, usually via the internet

URL (uniform resource locator) – the web site address. All web addresses are laid out in the same way

USB connection – a port which allows an ultra-fast download from a digital camera

Usenet – the network of newsgroups on the internet

VDU – visual display unit (the 'screen')

viewers – programs which allow you to see files on your computer

virus – a piece of software which enters your computer and has some, usually damaging, effect. Most viruses are written by malicious people

WAP (wireless application protocol) – programs which standardise internet communications so that they can be passed through computers, mobile phones, radios, etc.

web-based email – email which is associated with a web server

webmaster – the person who maintains a web site

web page – several web pages make up a web site

web site – text pictures and other kinds of information that form a unit of information on a subject. Web sites are usually made up of several linked web pages

webspace – the memory on a server in which a web site is located

welcome page – the first page you see when you visit a web site

wizard – a program which takes on a specific task to help you. For example, Chart Wizard in Excel helps you to draw the graph you want

world wide web (www) – this is a particular but major grouping of multimedia software programs and users on the net. There are parts of the internet which are not available on the world wide web

zip – a program which allows the compression of large files for transfer over the internet. Such files need to be unzipped before they can be used or read

References

Ausubel, D. P., Novak, J. D. and Hanesian, H. (1978) *Educational psychology: a cognitive view*. 2nd edition. Holt, Rinehart, and Winston, New York.

Barrass, R. (1990) Scientific writing for publication; a guide for beginners. *J. Bio. Ed.* **24**: 177–181.

BBC English Dictionary (1992) HarperCollins, London. 1374 pp.

Berners Lee, T. (2000) *Guardian*, 14 September, p. 7.

Booth, V. (1984) *Communicating in science. Writing and speaking*. Cambridge University Press, Cambridge. 68 pp.

Chellen, S. S. (2000) *The essential guide to the internet for health professionals*. Routledge, London. 215 pp.

Clarke, Arthur C. (2000) *The Observer*, 31 December p. 32.

Crumlish, C. (1999) *The internet for busy people. Millenium edition*. Osborne/McGraw-Hill, Berkeley, 264 pp.

Dussart, G. B. J. (1990) Creativity in the language of biology. *J. Bio. Ed.* **24**: 277–282.

Fishman, S. (1995) *Copyright your software*. Nolo Press, Berkely, Calif.

Frink, B. (1998) *Internet complete*. Sybex, San Francisco. 1021 pp.

Guttman, B. S. (1999) *Biology* WCB/McGraw-Hill, Boston. 1177 pp.

Kleiner, K. (1997) Publish on the net and be damned. *New Scientist*, 6 December, p. 16.

Lee, S., Groves, P. and Stephens, C. (1997) *Existing tools and projects for on-line teaching*. Report 005, Joint Information Systems Committee Technology Applications Programme, UK: Oxford University.

Li, X. and Crane, B. (1996) *Electronic styles. A guide to citing electronic information*. New Jersey, USA: Medford.

Lynas, C. (2000) *The eworld handbook*. Prentice Hall, Harlow.

Murray, P. D. F. (1952) *Biology; an introduction to medical and other studies*. Macmillan, London. 600 pp.

Naughton, J. (1999) *A brief history of the future. The origins of the internet*. Weidenfeld and Nicholson, London. 320 pp.

Roberts, M. B. V. (1972) *Biology: a functional approach*. Nelson, London. 626 pp.

Rodgers, J. and Schellenberg, K. (1996) *Computers in society* 6th edn. Dushkin Publishing Group, New York.

Royan, B. (2000) *Guardian Online*, 14 September, p. 4.

Sang, D. (2000) *Science @www Getting started – a resource for secondary schools*. Association for Science Education, Hatfield, Herts.

Schofield, J. (2000) *Guardian On-line*, 23 November.

Shaw, D. and Moore, H. (1996) *Electronic publishing in science*. Proceedings of a joint ICSU and UNESCO conference, Paris, 19–23 February 1996.

Thompson, J. (2001) IEEM Report to the British Ecological Society, *The Bulletin of the BES*, **32**: p. 17.

Wentk, R. (1998) *The Which? guide to the internet*. Which? Books, London.

Index